Online Decision Support for Offshore Wind Farm Installations

AF212548

Daniel Rippel

Online Decision Support for Offshore Wind Farm Installations

Springer Vieweg

Daniel Rippel
Fachbereich 4: Produktionstechnik
-Maschinenbau und Verfahrenstechnik-
Universität Bremen
Bremen, Germany

ISBN 978-3-658-49911-2 ISBN 978-3-658-49912-9 (eBook)
https://doi.org/10.1007/978-3-658-49912-9

© The Editor(s) (if applicable) and The Author(s), under exclusive license to Springer Fachmedien Wiesbaden GmbH, part of Springer Nature 2026

This work is subject to copyright. All rights are solely and exclusively licensed by the Publisher, whether the whole or part of the material is concerned, specifically the rights of translation, reprinting, reuse of illustrations, recitation, broadcasting, reproduction on microfilms or in any other physical way, and transmission or information storage and retrieval, electronic adaptation, computer software, or by similar or dissimilar methodology now known or hereafter developed.
The use of general descriptive names, registered names, trademarks, service marks, etc. in this publication does not imply, even in the absence of a specific statement, that such names are exempt from the relevant protective laws and regulations and therefore free for general use.
The publisher, the authors and the editors are safe to assume that the advice and information in this book are believed to be true and accurate at the date of publication. Neither the publisher nor the authors or the editors give a warranty, expressed or implied, with respect to the material contained herein or for any errors or omissions that may have been made. The publisher remains neutral with regard to jurisdictional claims in published maps and institutional affiliations.

This Springer Vieweg imprint is published by the registered company Springer Fachmedien Wiesbaden GmbH, part of Springer Nature.
The registered company address is: Abraham-Lincoln-Str. 46, 65189 Wiesbaden, Germany

If disposing of this product, please recycle the paper.

Acknowledgment

I wrote this thesis during my time as a research associate at BIBA—Bremer Institut für Produktion und Logistik GmbH at the University of Bremen. The topic resulted from the research project "OffshorePlan—Complementary application of mathematical and discrete-event models to solve complex planning and control problems in offshore construction logistics, which was funded by the German Research Foundation (DFG). In this context, I'd like to thank Dr.-Ing. Michael Lütjen and Professor Dr.-Ing. Helena Szczerbicka for the opportunity to work on their project and to write this thesis. Several persons and colleagues supported, whom I want to thank for their support and the lively discussions we had during this time.

My special thanks go to Professor Dr.-Ing. Michael Freitag, director of the BIBA and head of the research group Planning and Control of Production and Logistics Systems (PSPS) at the University Bremen, for supervising my thesis. Moreover, I'd like to thank Professor Dr. oec. Julia Arlinghaus for supporting me as my secondary reviewer with her constructive feedback.

This thesis would not have been possible without the discussions and constructive criticism from my colleagues. Among others, I want to thank Nicolas Jathe and Benjamin Staar, with whom I shared an office, for always having an open ear and the many inspirations they provided. Moreover, I want to thank Dr.-Ing Abderrahim Ait Alla, Stephan Oelker, Professor Dr.-Ing Thies Beinke, who, as specialists in the offshore wind area, always provided me with inputs and ideas for this thesis. I'd also like to thank Dr. Jürgen Pannek, Dr.-Ing. Haniyeh Dastyar, and Dr.-Ing. Tobias Sprodowski for lively discussions about this thesis's methodological background. Last but not least, I want to thank my parents and my wife for always supporting me. Without their support and love throughout my life, I would not have been where and who I am now.

Zusammenfassung

Entwicklung einer Online-Scheduling-Methodik zur Entscheidungsunterstützung für die Offshore-Windparkerrichtung unter Berücksichtigung von Wettervorhersageungenauigkeiten

Offshore-Windparks stellen eine vielversprechende Option zur Erzeugung grüner, nachhaltiger Energie dar. Im letzten Jahrzehnt hat die installierte Offshore-Kapazität weltweit exponentiell zugenommen. Dieser Anstieg resultiert aus der stetigen Weiterentwicklung zu größeren und leistungsfähigeren Turbinen, aber auch aus einer konstanten Zunahme der Anzahl und Größe von Offshore-Windparks. Die Literatur gibt an, dass etwa 15%–30% der Kosten eines Offshore-Windparks auf die Logistik während der Errichtung entfallen. Diese Kosten resultieren hauptsächlich aus der Wetterabhängigkeit des Installationsprozesses, welche die Planung teurer Ressourcen, z. B. von Installationsschiffen oder Hafenausrüstung erschwert. Diese Abhängigkeit beeinflusst die Planung auf allen Ebenen, indem sie starke Unsicherheiten induziert. Die operative Ebene ist mit der Unsicherheit von Wettervorhersagen und volatilen Wetterbedingungen auf See konfrontiert. Darüber hinaus müssen Planer auf taktischer und strategischer Ebene auf Erfahrungen, Schätzungen und historische Daten zurückgreifen. In der aktuellen Literatur finden sich einige Modelle zur Unterstützung bei strategischen oder taktischen Planungsaufgaben. Jedoch existieren keine Ansätze auf operativer Ebene, da diese nicht mit aktuellen Wetterdaten und -vorhersagen umgehen können. Entsprechend wird in dieser Arbeit eine dedizierte Online-Planungsmethodik zur Entscheidungsunterstützung vorgestellt, die explizit darauf ausgelegt ist, aktuelle Wettermessungen und –vorhersagen einzubeziehen. Die vorliegende Arbeit bettet diese Methodik in ein Entscheidungsunterstützungssystem ein, das in der Lage ist, Entscheidungsunterstützung für verschiedene Planungsaufgaben auf allen drei Ebenen zu bieten.

Das Entscheidungsunterstützungssystem verwendet ein Model Predictive Control-Schema in Kombination mit einem Mixed-Integer-Linear-Programming Optimierungsmodell, um eine prädiktiv-reaktive Planung des Installationsprozesses zu erreichen. In Erweiterung der aktuellen Literatur erzeugt diese Methodik kurz-bis mittelfristige Zeitpläne für jedes Schiff, anstatt direkte Steuerungsentscheidungen zu generieren. Um eine solche Planung zu ermöglichen, verwendet die Methodik eine neuartige Methode, welche Prognoseunsicherheiten einbezieht. Diese Methode bettet zunächst die Vorhersageunsicherheiten in die Vorhersagedaten ein und wandelt diese Daten damit in Wahrscheinlichkeitsintervalle für mögliche Wetterbedingungen um. Danach wendet sie so genannte Laufzeitschätzer an, die versuchen, den Einfluss der Intervalle auf die Dauer von Offshore-Operationen vorherzusagen. Schließlich dienen die geschätzten Dauern als Eingabe für ein angepasstes Mixed-Integer-Optimierungsmodell, um optimale Pläne zu berechnen. Die Methodik bettet diesen Prozess in das Modell-prädiktive Steuerungsschema ein, um einerseits den Umfang des Optimierungsproblems zu reduzieren und andererseits, um die notwendige Reaktivität zu erreichen. Dieses Schema besteht aus einer Kombination von geschlossenen und offenen Regelkreisen, die eine einzigartige Kombination von prädiktivem und reaktivem Verhalten ermöglichen. Die Integration von Vorhersagen und deren Unsicherheit ermöglicht prädiktive Optimierungen für längere Zeiträume im offenen Regelkreis. Danach stellt das Schema in seinem geschlossenen Regelkreis den ersten Planabschnitt für das reale System (Installationsprozess) bereit, bevor es die Situation neu bewertet und zu einer neuen offenen Optimierung übergeht. Folglich bietet diese Methodik einen passenden Ansatz zur Entscheidungsunterstützung auf operativer Ebene, liefert aber auch eine Grundlage für Offline-Optimierungsaufgaben auf höheren Planungsebenen. Daher wird in dieser Arbeit die Verwendung einer Simulationskomponente als Teil des Entscheidungsunterstützungssystems vorgeschlagen. Durch die Simulation von Projekten unter Verwendung von aggregierten historischer Daten ermöglicht die Methodik eine Abschätzung operativer Informationen, z. B., für die Planung von Kapazitäten oder Transportzyklen.

Die Evaluation des Entscheidungsunterstützungssystems, der ihm zugrunde liegenden Methodik und der verschiedenen Komponenten zeigt ein hohes Potenzial für Kosteneinsparungen auf operativer Ebene. Ein Benchmark mit dem aktuellen Stand der Praxis zeigt eine Reduktion der variablen Kosten um ca.

9%. Darüber hinaus zeigt der vorgeschlagene prädiktiv-reaktive Ansatz, deutliche Vorteile gegenüber dem aktuellen Stand der Forschung, d. h. gegenüber robusten oder regelungsbasierten Ansätzen. Der vorgeschlagene Ansatz liefert zuverlässige Pläne für etwa eine Woche. Diese Zeit hängt dabei von der verfügbaren Optimierungszeit und den Unsicherheiten der Wetterprognosen ab. Diese Zuverlässigkeit resultiert in weniger als 2% zusätzlicher Kosten im Vergleich zu einer regelungsbasierten Annahme. Im Gegensatz dazu verringert der prädiktiv-reaktive Ansatz die Planzuverlässigkeit nur geringfügig, senkt jedoch die Kosten um mehr als 1,3% im Vergleich zu einer robusten Grundannahme. Schließlich zeigt die Evaluation die Eignung der vorgeschlagenen Methodik zur Unterstützung von Offline-Planungsaufgaben. Hierbei kann die Methodik direkt mit aggregierten historischer Daten umgehen, um verlässliche Schätzungen für die Anzahl der installierbaren Turbinen zu liefern.

Zusammenfassend zeigen die Ergebnisse, dass die Kombination eines modellprädiktiven Schemas mit einem Mixed-Integer-Optimierungsmodell eine hervorragende Grundlage für die Entscheidungsunterstützung auf operativer Ebene bietet. Die vorgeschlagene Methode zur Integration und Handhabung von Vorhersageunsicherheiten stellt eine gute Möglichkeit dar, um die Effizienz von Planungsaufgaben im Offshore-Bereich zu steigern.

Abstract

Development of an Online Scheduling Methodology for Decision Support for Offshore Wind Farm Installations under Consideration of Weather Forecast Uncertainties

Offshore wind farms constitute a promising option to generate green, sustainable energy. The last decade has witnessed an exponential increase in installed offshore capacity world-wide. This increase resulted from steady advancements to more extensive and capable turbines and witnessed a constant increase in offshore wind farms' number and size. Literature attributes that approximately 15%–30% of an offshore wind farm's logistics costs during the construction. These costs mainly result from the installation process' weather dependency, which renders the planning of expensive resources like jack-up vessels or port-side equipment complicated. This dependency influences planning on all levels. The operative level faces the uncertainty of weather forecasts and volatile weather conditions at sea. Moreover, planners need to rely on experience, estimates, and historical data on the tactical and strategic levels. Current literature proposes some models to support planners with higher-level strategic or tactical planning tasks but does not offer any support on the operative level. Consequently, this thesis presents a dedicated online scheduling methodology for decision support, explicitly designed to incorporate current weather measurements and forecasts. This thesis embeds this methodology into a decision support system, capable of providing decision support for various planning tasks on all three levels.

The decision support system uses a Model Predictive Control scheme combined with a Mixed-Integer Linear Programming optimization model to achieve predictive-reactive scheduling of the installation process. This methodology obtains short- to mid-term schedules for each vessel instead of obtaining control decisions, extending current literature. Therefore, the methodology applies a newly proposed method to include forecast uncertainties. This method first encodes the forecast uncertainty into the forecast data, effectively turning this data into a set of intervals for probable weather conditions. Afterward, it applies so-called duration estimators to predict the intervals' influence on the offshore operations' duration. Finally, it applies a custom Mixed-Integer optimization model to calculate optimal plans using these estimated durations. The methodology embeds this process into a model predictive control scheme to reduce the optimization problem's size and achieve reactivity. This control scheme consists of a combination of closed-and open-loop controls, which enable a unique combination of predictive and reactive features. The integration of forecasts and their uncertainty allows predictive open-loop optimizations for prolonged time frames. Afterward, the scheme applies the first segment of these plans to the real-world system in its closed-loop before reevaluating the situation and progressing to a new open-loop. Consequently, this methodology offers a viable tool for decision support on the operative level and provides a foundation for offline optimization tasks on higher planning levels. Therefore, this thesis proposes the use of a simulation component as part of the decision support system. By simulating projects using aggregates of historical data, the methodology allows an estimation of operative information, e.g., for the planning of capacities or transport cycles.

The evaluation of the decision support system, its underlying methodology, and its different components show high potential for cost savings on the operative level. A benchmark with the current state of practice shows a reduction in variable costs of approximately 9%. Moreover, the predictive-reactive approach implemented by this decision support system shows distinctive advantages over current state of the art methods, i.e., robust or control-based approaches. The approach delivers reliable plans for approximately one week, depending on the available optimization time and forecast uncertainty. This reliability comes at less than 2% of additional costs compared to a control-based assumption. In contrast, the predictive-reactive approach only slightly reduces the plan-reliability but reduces costs by more than 1.3% compared to a robust assumption. Finally, the evaluation demonstrates the feasibility of the proposed methodology to support offline planning tasks. The methodology can directly handle aggregates of historical data to provide good estimates for the number of installable turbines.

In conclusion, the results show that combining a Model Predictive Control scheme with a Mixed-Integer scheduling model provides a feasible foundation to provide decision support on an operative level. The proposed method to integrate and handle forecast uncertainty constitutes a viable way to enhance planning and scheduling tasks in the offshore area.

Contents

Abbreviations

DFG	German Research Foundation (Deutsche Forschungsgemeinschaft)
DSRM	Design Science Research Methodology
DSS	Decision Support System
EMF	Eclipse Modelling Framework
EMOF	Essential Meta-Object Facility
GIS	Geographic Information System
HLV	Heavy-lift Vessel
JSSP	Job-shop Scheduling Problem
M2M	Model-To-Model
M2T	Model-To-Text
MAS	Multi-agent Simulation
MDD	Model Driven Development
MILP	Mixed-Integer Linear Program
MPC	Model Predictive Control
OCL	Object Constraint Language
OMG	Object Management Group
OWF	Offshore Wind Farm
OWT	Offshore Wind Turbine
RCP	Eclipse Rich Client Platform
T2M	Text-To-Model
TSP	Traveling Salesmen Problem
UML	Unified Modeling Language

Nomenclature

Symbol	Domain	Unit	Description
Indices			
i	$\mathbb{N}^+$	Index	Running index
t_i	$\mathbb{N}^+$	Index	i-th sampling instance
p_i	$\mathbb{N}^+$	Index	i-th planning period
v	$\mathbb{N}^+$	Index	v-th vessel in V
k	$\mathbb{N}^+$	Index	k-th time instance usually in N
o	$\mathbb{N}^+$	Index	Index of operations for the scheduling model with 1 = install, 2 = move to port, 3 = move to site, and 4 = load components
r	$\mathbb{N}^+$	Index	Index for possible Routes during transport planning
c	$\mathbb{N}^+$	Index	Index for components / production ports, usually as 1 = tower, 2 = blades, 3 = nacelle
j	$\mathbb{N}^+$	Index	j-th Round-trip within a resupply cycle
l	$\mathbb{N}^+$	Index	l-th resupply cycle within a project
Parameters			
T	$\mathbb{N}^+$	Time Units	Sampling step width
P	$\mathbb{N}^+$	–	Number of planning periods
N	$\mathbb{N}^+$	Time Units	Prediction horizon

(continued)

(continued)

Symbol	Domain	Unit	Description
$D_{o,k}$	$\mathbb{N}^+$	Time Units	Estimated duration of operation o at time instance k
$x(t)$	$\mathbb{R}$	Vector	State at sampling instance t
$x_*(t)$	$\mathbb{R}$	Vector	Reference signal for sampling instance t
$x_{u^\star}(t)$	$\mathbb{R}$	Vector	Sequence of predicted states for sampling instance t and optimal control sequence $u^\star$
$u_*(t)$	$\mathbb{R}$	–	Reference control for sampling instance t
λ	$\mathbb{R}$	–	Weighting factor
N^{owt}	$\mathbb{N}^+$	–	Number of turbines to install
$CUR_{t_i}^{owt}$	$\mathbb{N}^+$	–	Number of turbines finished at sampling instance t_i
V	$\mathbb{N}^+$	–	Number of installation vessels
$BAYS$	$\mathbb{N}^+$	–	Number of loading bays at the base port
CAP^{port}	$\mathbb{N}^+$	Comp. Sets	Maximum port capacity to store component sets
CAP_v	$\mathbb{N}^+$	Comp. Sets	Maximum capacity of vessel v to store component sets
RES^{frq}	$\mathbb{N}^+$	hours	Minimum time required for base port a resupply cycle
RES^{amt}	$\mathbb{N}$	Comp. Sets	Number of complete component sets delivered in a single resupply cycle
$RES_{t_i}^{earliest}$	$\mathbb{N}^+$	Time Units	Minimum number of hours until the base port can receive another shipment of components
$RES^{initial}$	$\mathbb{N}^+$	Component Sets	Initial storage of components at the base port
INV^-	$\mathbb{N}_0^+$	Component Sets	Minimum desired inventory during capacity optimization
N^c	$\mathbb{N}^+$	–	Number of component types / production ports
N^{rc}	$\mathbb{N}^+$	–	Number of resupply cycles during an installation
N^{rt}	$\mathbb{N}^+$	–	Number of round-trips within a resupply cycle
N^{years}	$\mathbb{N}^+$	years	Number of historical years used in the capacity optimization

(continued)

(continued)

Symbol	Domain	Unit	Description
$PLAN_v^{vessel}$	$\mathbb{N}^+$	–	Index of time instance when a vessel v is available
$PLAN_k^{bays}$	$\mathbb{N}$	–	Denotes the number of already scheduled busy loading bays for each time instance k
$PLAN_k^{load}$	$\mathbb{N}$	–	Denotes the number of already scheduled loading operations started at each time instance k within the prediction horizon
$C_v^{offshore}$	$\mathbb{R}$	€/hour	Cost for vessel v for being offshore
$C_v^{loading}$	$\mathbb{R}$	€/hour	Cost for vessel v for loading operations
C_v^{fuel}	$\mathbb{R}$	€/hour	Fuel cost for vessel v for movement
B^{owt}	$\mathbb{R}$	(€)	Optimizer benefit for finishing an installation
B^{early}	$\mathbb{R}$	(€)	Optimizer benefit for finishing an operation early
B^{inv}	$\mathbb{R}$	(€)	Optimizer penalty for high inventory levels
D_c^{setup}	$\mathbb{R}$	hours	Setup cost for component c
D_c^{load}	$\mathbb{R}$	hours	Loading cost for component c
D_c^{unload}	$\mathbb{R}$	hours	Unloading cost for component c
D_r^{route}	$\mathbb{R}$	hours	Movement cost for route r
$CAP^{payload}$	$\mathbb{R}^+$	tons	Maximum payload for HLVs
CAP^{area}	$\mathbb{R}^+$	m^2	Maximum deck area for HLVs
$REQ_c^{payload}$	$\mathbb{R}^+$	tons	Weight of component c
REQ_c^{area}	$\mathbb{R}^+$	m^2	Required deck area of component c
$\delta(k)$	$\mathbb{R}^+ \in [0..1]$	$\frac{\%}{100}$	Uncertainty estimation
$forecast_s(k)$	$\mathbb{R}^+$	m/s	Lower and upper bound for wind speeds at instance k
$forecast_h(k)$	$\mathbb{R}^+$	m/s	Lower and upper bound for wave heights at instance k
$\mu_s(k)$	$\mathbb{R}^+$	m/s	Forecasts - Expected wind speed at instance k
$\mu_h(k)$	$\mathbb{R}^+$	m	Forecasts - Expected wave height at instance k
μ_s^*	$\mathbb{R}^+$	m/s	Forecasts - Average wind speed over N
μ_h^*	$\mathbb{R}^+$	m	Forecasts - Average wave height over N
Variables			
$u(t)$	$\mathbb{R}$	–	Control / feedback for sampling instance t

(continued)

(continued)

Symbol	Domain	Unit	Description
$u^{\star}(t)$	$\mathbb{R}$	–	Optimal control sequence for sampling instance t
Optimizer Decision Variables			
$Y_{v,k}^{owt}$	Binary	–	Denotes if vessel v starts an installation at time instance k
$Y_{v,k}^{toSite}$	Binary	–	Denotes if vessel v starts moving to the installation site at time instance k
$Y_{v,k}^{toPort}$	Binary	–	Denotes if vessel v starts moving to the port at time instance k
$Y_{v,k}^{loading}$	Binary	–	Denotes if vessel v starts a loading operation at time instance k
$Y_{k}^{resupply}$	Binary	–	Denotes if the base port receive a resupply at time instance k
$X_{j,c}^{freight}$	$\mathbb{N}^{+}$	Component	Number of components of type c transported in round-trip j
$X^{capacity}$	$\mathbb{N}^{+}$	Comp. Sets	Base port capacity during capacity optimization. Later as parameter CAP^{port}
$X^{initial}$	$\mathbb{N}^{+}$	Comp. Sets	Base port initial inventory level during capacity optimization. Later as parameter $RES^{initial}$
Optimizer Dependent Variables			
$Y_{v,k}^{loc}$	Binary	–	Denotes if vessel v is in port (0) or offshore (1) at time instance k
$Y_{v,k}^{atBay}$	Binary	–	Denotes if vessel v occupies a loading bay (1) or not (0) at time instance k
$Y_{v,k}^{busy}$	Binary	–	Denotes if vessel v is currently performing an operation (1) or is idle (0)
X_{k}^{invP}	$\mathbb{N}^{+}$	–	Denotes the current amount of component sets at the base port at time instance k
$X_{v,k}^{invV}$	$\mathbb{N}^{+}$	–	Denotes the current amount of component sets stored in vessel v at time instance k
$X_{v,k}^{fOwt}$	$\mathbb{N}^{+}$	–	Denotes at which time instance an installation operation conducted by vessel v and started at instance k will finish

(continued)

(continued)

Symbol	Domain	Unit	Description
$X_{v,k}^{fOp}$	$\mathbb{N}^{+}$	–	Denotes at which time instance a non-installation operation conducted by vessel v and started at instance k will finish
$X_{j,c}^{setup}$	$\mathbb{N}^{+}$	Component	Number of setup operations required in round-trip j for each component type c
$Y_{j,c}^{port}$	Binary	–	1 if the HLV transports components of type c in the j-th round-trip; 0 otherwise
$Y_{j,r}^{route}$	Binary	–	1 if route r describes the visited production ports for round-trip j; o otherwise
$X_{l}^{invCurrent}$	$\mathbb{N}^{+}$	Comp. Sets	Inventory level at time period l

List of Figures

List of Tables

Introduction 1

This chapter briefly introduces this thesis by summarizing its motivation, goals, and research questions. While these parts mainly constitute summaries of later chapters' results, this introduction also presents the overall research methodology, the structure of this thesis, and a list and discussion of previously published research that directly ties into this thesis.

1.1 Motivation

Over the last decades, wind energy has evolved into a prominent contributor to sustainable, green energy. The amount of energy produced by wind farms increased exponentially over the last ten years, for example, resulting in an increase of about 11% (52 Gigawatt) to a total of 539 Gigawatt worldwide in 2017 (REN21, 2019). During the first half of 2019, Germany installed approximately 1350 new offshore wind turbines (OWTs) with a cumulative performance of 6.6 Gigawatt (Deutsche WindGuard GmbH, 2019).

In the context of wind energy, research and practice distinguish between onshore and offshore farms. While both types of wind farms require the same components and processes, their underlying characteristics and legal frames differ. Thereby, (OWFs) produce higher amounts of energy due to the higher availability and strength of wind at the open sea (Breton and Moe, 2009). Moreover, the open sea offers more space to construct OWFs than the mainland. For example, Germany currently debates the minimum distance between a wind turbine and the nearest settlement. Current regulations reach from 300 meters to up to 1,100 meters for larger settlements or even single houses (Fachagentur Windenergie an Land e.V., 2020).

Although OWFs provide several advantages, their construction poses additional challenges compared to their onshore counterparts, e.g., summarized in Rippel et al.

© The Author(s), under exclusive license to Springer Fachmedien Wiesbaden GmbH, part of Springer Nature 2026
D. Rippel, *Online Decision Support for Offshore Wind Farm Installations*,
https://doi.org/10.1007/978-3-658-49912-9_1

(2019a); Schweizer et al. (2011). In particular, dynamic weather conditions at sea, coupled with specialized and expensive resources, result in extensive costs during the installation. Literature attributes between 15% and 30% of an OWF's total costs to logistics during the installation (Dewan et al., 2015; Lange et al., 2012; Muhabie et al., 2018). For example, installation vessels, so-called jack-up vessels, attributed charter costs of up to 145,000€ per day in 2014 (Meyer, 2014). As the number of installation projects continuously increases (c.f. Deutsche WindGuard GmbH (2019) for an example from Germany) and will probably increase due to additional decommissioning projects for old OWFs Beinke et al., 2020a, the prices probably remain this high or increase further over the next years. Current newspaper articles support this assumption. For example, Durakovic (2020) describes a study that expects the demand for offshore installation vessels to increase from an annual rate of 13 vessel years in 2020 to over 40 vessel years by 2025. In turn, the article states that world-wide, a total of 32 installation vessels exists, with five more ordered or in production. Moreover, it shows that only a subset of these vessels can handle newer generation turbines due to their increasing size and weight. In addition to vessels, port-side resources like heavy-duty storage areas or handling equipment might evolve into expensive bottlenecks (Oelker et al., 2020).

This combination of high costs for chartered resources and unstable weather conditions at the installation sites pose project planners with a particular problem. Although they need to plan operations and resources precisely, highly dynamic weather conditions render reliable long-term planning difficult to impossible. In this context, planning can only rely on historical information and current weather forecasts, each imposing a high degree of uncertainty. In particular, offshore operations scheduling plays a vital role in efficient project planning and execution. While online scheduling allows efficient use of resources, offline scheduling can support the mid- to long-term planning. In practice, port operatives, project managers, and the vessels' captains generally render their scheduling decisions ad-hoc, based on the current situation and available weather forecasts (Rippel et al., 2019b). Thus, it can be assumed that a suitable decision support system might positively impact the online scheduling of offshore operations.

1.2 Research Aim

This research aims to develop a methodology that acts as a decision support system (DSS) for scheduling offshore operations during wind farm installations. This methodology's development focuses on three main functional aspects, as shown in Fig. 1.1.

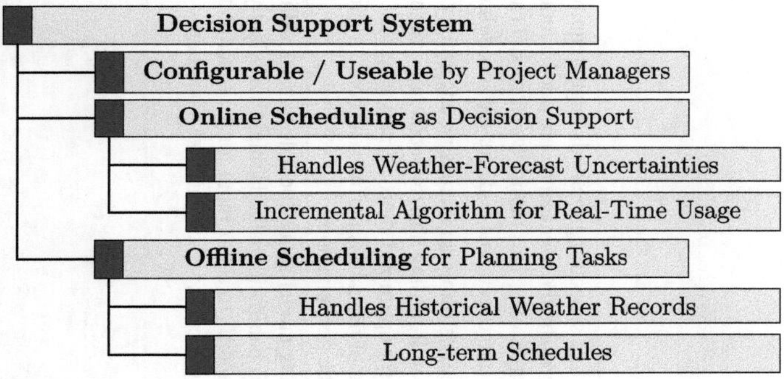

Fig. 1.1 Main functional requirements towards the methodology

The methodology focuses on usage by project managers and port officials. Thus, it requires methods for detailed but easy to use *customization*. After the methodology's initial setup, project managers should provide the project's details, e.g., locations, vessel information, or processes, in a user-friendly way. Moreover, the methodology should support online and offline scheduling. This research considers *online scheduling* as a process-accompanying optimization. While the installation progresses or new weather forecasts become available, the methodology should derive updated plans as part of the DSS. Moreover, this research considers *offline scheduling* to use aggregates of historical weather data, e.g., for capacity or transportation planning, as project managers need to render such decisions long before current data becomes available. While online scheduling aims at providing short- to mid-term plans, offline scheduling uses historical information to estimate, e.g., the project duration or activities that could have taken place in a specific time window. Table 1.1 summarizes the research hypotheses and planned methods that this research applies and evaluates to realize these functionalities.

Besides developing the stated methodology, this thesis aims to provide answers to the following research questions. These questions follow from the hypotheses stated in Table 1.1 and the literature review presented later on in Chap. 4:

Table 1.1 Main functionalities, planned methods, and hypotheses

Functionality	Method(s)	Hypothesis
Customization	Meta-Modeling; Model Transformation; Code Generation	While the actual scheduling methods might require complex data structures in specific formats, the user can be provided a more suitable editor to define the project intuitively. This intuitive user model can automatically be transformed into executable code for the scheduling algorithm.
Online Scheduling	Mixed-Integer Linear Programming; Model Predictive Control; Forecast Discretization	Weather forecasts imply an increasing uncertainty the longer the prediction horizon grows. Receding horizon techniques might render this uncertainty manageable for optimized short- to mid-term plans. Methods for the discretization of probabilistic forecasts could avoid extensive runtimes of, e.g., stochastic optimizations.
Offline Scheduling	Historical Weather Data; Probabilistic Methods for Data Aggregation; Use-Case specific Optimization Models	The scheduling model should be able to work offline using historical data. Therefore, probabilistic methods might sufficiently aggregate multiple years of weather records to act as input for the scheduling. Additionally, several use-cases might exist, which can take advantage of (averaged) scheduling information.

RQ1 Is it possible to apply Model Predictive Control to the online scheduling of offshore operations in a planning context (not just control)? Which methods and models does it require to act as DSS?

RQ2 Does a combination of receding horizon techniques, particularly Model Predictive Control and Mixed-Integer Linear Programming scheduling models, provide advantages over classical scheduling approaches, e.g., stochastic or robust optimizations?

RQ3 Is it possible to apply the proposed scheduling algorithm to historical weather records to achieve suitable approximations for long-term planning tasks? Which additional preprocessing does such an application require?

RQ4 To which extend is it possible to use model transformation and code generation to provide project managers with a detailed and intuitive interface for the mathematical scheduling algorithm? Can other planning tasks be integrated into such an interface?

1.3 Research Methodology and Structure of the Thesis

This research uses the Design Science Research Methodology (DSRM) to structure research activities and this thesis. The DSRM originates from business information systems, which constitutes a crossover discipline between computer sciences, engineering, and business economics and was first proposed by Peffers et al. (2007). According to the authors, the DSRM aims to develop so-called artifacts to improve or solve a specific problem, which differentiates it from explanatory research. Therefore, the authors state that this methodology has proven particularly suitable for the engineering domain, leading scientists to follow a balanced theoretical and practical development process. Artifacts, in terms of the DSRM, range from but are not limited to algorithms, models, computer systems, or human-computer interfaces. In this context, the design and development of a DSS for scheduling offshore operations constitute the primary artifact. This development requires selecting or developing suitable components, e.g., a formulation for the scheduling or methods to handle forecasting uncertainties.

In literature, several Design Science Research or general engineering procedure models exist. This thesis chose the DSRM as it combines both a scientific approach with the artifact's practical development instead of just focusing on one or the other aspect. Most of the existing procedure models follow a similar general markup, starting with a situational analysis and progressing over the definition of objectives to the development, evaluation, and implementation. Thereby, each model imposes a different focus. For example, the Systems Engineering procedure model presented in Krallmann et al. (2013) emphasizes the actual realization and implementation of the developed tools within a company. Consequently, it does not explicitly cover scientific tasks, like dissemination. Other classical engineering or computer science procedure models have a similar focus, e.g., V- or Waterfall-models. Others follow a different structure, e.g., the systems engineering procedure model proposed by Daenzer and Haberfellner (2002) focuses on deriving alternative solutions for varying levels of detail and on their selection. Nevertheless, most of these models highly focus on the actual development and implementation. In contrast, Design Science Research procedure models put a stronger focus on scientific aspects of the development process and do not require an explicit implementation of the solution in a company or similar. For example, Hevner (2007) proposes a Design Science Research procedure model consisting of three cycles. The relevance cycle connects the design stage with the environment. It continuously delivers new requirements, and vice versa aims to evaluate the designed artifact within this environment, e.g., using field tests or workshops. The design cycle continuously improves and evaluates the artifact. Finally, the rigor cycle connects the development to the knowledge base by continuously providing new knowledge, e.g., publications, but also drawing from the knowledge base in terms of existing methods or approaches. While this procedure model integrates both the scientific and practical aspects very well, its highly iterative character seems unfitting for this thesis's topic. This procedure model seems mostly applicable to situations that require agile or iterative development approaches. Consequently, this thesis applies the procedure model proposed by Peffers et al. (2007), which shows a linear process that includes several feedback loops. Figure 1.2 depicts this procedure model and maps the single steps to this thesis' chapters.

This process model follows the general structure noted above, covering the situational analysis, objectives definition, development, demonstration, and evaluation. Additionally, this procedure model requires researchers to communicate and discuss their results, e.g., by publishing their research.

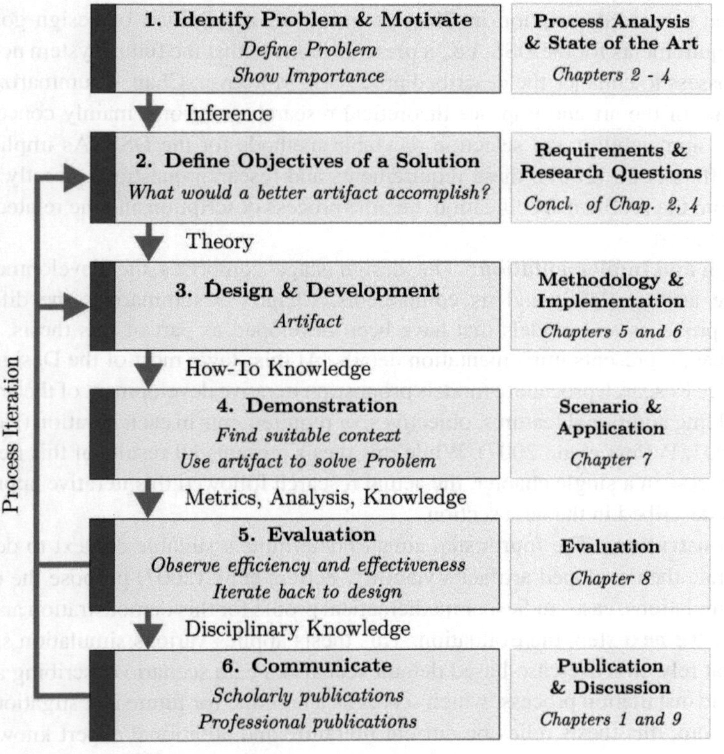

Fig. 1.2 Design Science Research Methodology. (Adapted from Peffers et al. (2007))

Identify problem and motivate: This first step aims to describe the problem and the respective domain. This thesis separates this step into two different areas. First, it describes the application domain, involved processes, resources, and restrictions in Chap. 2. Second, this thesis presents the state of the art, both regarding possibly applicable methods (Chap. 3) and existing approaches within the application domain (Chap. 4). This thesis relies on a broad literature review to compile these situational analyses. Additionally, the process description includes information obtained from other researchers that have conducted several research projects in the offshore area.

Define objectives of a solution: Both of these descriptions result in a deduction of objectives for the DSS or more theoretical research questions for this thesis.

The process description in Chap. 2 ends with a definition of design goals or requirements for the DSS, i.e., it presents features that the future system needs to possess to enhance the described processes. Moreover, Chap. 4 summarizes the state of the art and imposes theoretical research questions, mainly concerning the applicability and selection of viable methods for the DSS. As implied by Peffers et al. (2007), these requirements and research questions directly result from the problem specification, i.e., the process description and the related state of the art.

Design and implementation: The design stage comprises the development of the artifact (DSS) and its components. Chapter 5 summarizes the different approaches and models that have been developed as part of this thesis, while Chap. 6 presents implementation details. At this stage, most of the Design Science Research procedure models propose an iterative development of the artifact, adding additional features, objectives, or requirements in each iteration (Hevner, 2007; Peffers et al., 2007). While this thesis presents all results of this iterative process in a single chapter, the actual research followed this iterative approach, as described in the next section.

Demonstration: The fourth step aims to determine a suitable context to demonstrate the developed artifact's viability. Peffers et al. (2007) propose the use of simulations, case studies, or mathematical proofs for this demonstration and also for the next step, the evaluation. This thesis applies various simulation studies that rely on a use-case-based default scenario, i.e., a scenario describing a realistic installation process, which serves as a baseline for future investigations. As before, this thesis relies on current literature and additional expert knowledge to construct this default scenario. Moreover, Chap. 7 shortly describes a simulation run of the proposed DSS and its components for the default scenario. This demonstration shows how the DSS solves this problem. After this short demonstration, Chap. 7 introduces the experimental setup for the next step, identifying aspects that need further investigation to allow a final evaluation of the imposed requirements.

Evaluation: The fifth step aims to provide an in-depth understanding of the artifact's efficiency and effectiveness. As proposed by Peffers et al. (2007), this thesis presents several simulation studies in Chap. 8, ranging from evaluations of different ways to handle historical weather data over benchmarks to exemplary DSS applications for the optimization transport routes or the base port capacity.

Communication: The final stage of the procedure model requires scientists to publish and communicate their work. Therefore, a critical discussion of the research and its results constitutes a crucial part of such dissemination. Chapter 9 presents such a discussion, reflecting the presented results in the context of the design goals and research questions. Finally, Chap. 9 presents various impulses for future work. Several research articles have been published as part of the research leading to this thesis. The following section lists these publications and shows that the research, while following this procedure model, took several iterations, each adding new aspects to the DSS's design.

1.4 Previous Work and Publications

The topic of this thesis has resulted from the research project *OffshorePlan—Complementary application of mathematical and discrete-event models to solve complex planning and control problems in offshore construction logistics* that was funded by the German Research Foundation (Deutsche Forschungsgemeinschaft) (DFG) under the grant number (LU 2049/1-1 | SZ 51/33-1). In consequence, several aspects of this thesis have already been published in various conference and journal articles. An overview of these articles and the related aspects of this thesis is given in Table 1.2.

Table 1.2 Previously published articles using content from this thesis

Reference	Conference/Journal	Related Content
Rippel et al. (2019a)	IFAC-Papers Online	State of the art & process description (literature review)
Rippel et al. (2019d)	WinterSim'19	First version of the online scheduling methodology (proof-of-concept)
Rippel et al. (2019b)	Applied Sciences	Extended online scheduling and evaluation
Rippel et al. (2020a)	SummerSim'20	Offline application: capacity optimization
Rippel et al. (2020b)	WinterSim'20	Offline application: transport optimization
Rippel et al. (2019c)	HICL'19	Meta-model for offshore scheduling
Rippel et al. (2020c)	HICL'20	Model transformation framework & dedicated user interface

Referring back to the research methodology, these publications' order and content demonstrate how the procedure model was applied. The first article focused on describing the area of application (wind farm installations), processes, requirements, and the state of the art in this domain. The article concluded with a deduction of planning tasks that would benefit from a DSS and proposed methods that might be applied (Rippel et al., 2019a). Afterward, Rippel et al. (2019d) presented an early version of the online scheduling methodology and demonstrated its feasibility. Rippel et al. (2019b) finally extended the methodology to cover additional restrictions already identified within the first article, e.g., port-side resources. Referring back to the procedure model, this constitutes an example for the iterative loop from the last step to the objectives step: The second article demonstrated that the proposed methodology works for a subset of the identified problems. Thus, the next iteration widened the objectives to cover additional restrictions and tasks. Similarly, Rippel et al. (2020a,b) extended the scheduling methodology to evaluate its feasibility for additional offline planning tasks, i.e., planning transport routes or the base port capacities. Therefore, these articles proposed additional algorithms and models for these tasks. Finally, Rippel et al. (2019c, 2020c) incrementally proposed a meta-model and model transformation framework that allows separating the DSS's user interface from the actual models and methods.

While this thesis summarizes all these models and concepts in a single chapter, the actual research followed the implied iterative approach, continuously including additional tasks or restrictions.

Process Description and Requirements 2

This chapter introduces the general process for the installation of OWFs. Section 2.1 presents general trends and developments within the wind energy sector. These trends show a steady, worldwide increase in the production of renewable energy and, in particular, in offshore wind energy, resulting in an increasing share of renewable energy in worldwide energy production. Afterward, Sect. 2.2 presents the installation process and a brief description of the required resources and constraints. Section 2.3 provides the baseline for decomposing the overall planning problem into smaller, more focused planning tasks with different planning horizons. Finally, Sect. 2.4 describes the focus and limitations of this thesis, i.e., its focus on scheduling offshore operations during the installation of top structures. Moreover, it describes the deduction of requirements towards the methodology and the DSS presented in this thesis. These requirements provide a foundation for evaluating the state of the art and justify selected methods in the following Chap. 3.

2.1 Development of the Wind Energy Sector

Over the last decade, wind energy has received much attention in research and practice. This increase in attention results from the steady increase in renewable energy worldwide. According to the Renewables 2019 Status Report, 162 countries had acts or laws to increase their share of renewable energy to percentages ranging from 20% to 100% of their energy consumption by the end of 2018 (REN21, 2020). For example, in 2020, the European Union just raised its target share from 27% to 32% by the end of 2030. Germany's 2021 Renewable Energy Sources Act aims at shares of 65% of the gross electricity consumption by 2030 and at a minimum of 80% by 2050 (EEG, 2021). In particular, Germany aims to increase the installed offshore capacity from approximately 7.76 Gigawatt by 2020 (Stiftung

© The Author(s), under exclusive license to Springer Fachmedien Wiesbaden GmbH, part of Springer Nature 2026
D. Rippel, *Online Decision Support for Offshore Wind Farm Installations*,
https://doi.org/10.1007/978-3-658-49912-9_2

OFFSHORE-WINDENERGIE, 2020) to at least 20 Gigawatt by 2030 and further to at least 40 Gigawatt by 2040 (WindSeeG, 2020).

2.1.1 Installed Capacity

Like the International Renewable Energy Agency (IRENA) or REN21, several organizations track the development of renewable energy worldwide. Figure 2.1 depicts the annual installed capacity for wind energy in general (left axis) and specifically for offshore wind farms (right axis). In detail, Fig. 2.1a shows the values obtained from IRENA (Whiteman et al., 2020), while Fig. 2.1b depicts the values provided by REN21 (REN21, 2020).

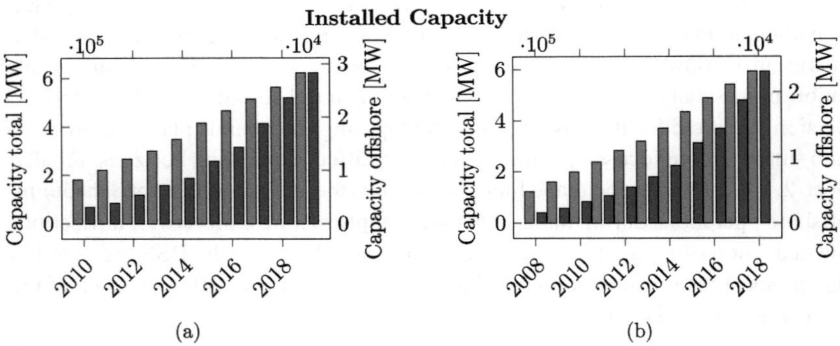

Fig. 2.1 Annual values for the total and offshore wind energy capacity worldwide. (Sources: (a) Whiteman et al. (2020) (b) REN21 (2020))

In absolute numbers, REN21 (2019) describes that in 2017, new wind farms producing approximately 52 Gigawatts were constructed worldwide, increasing the total amount of energy produced by wind turbines by approximately 11% to a total of 539 Gigawatts (REN21, 2019). In 2018, their updated report states an increase of an additional 51 GW for a total of 591 GW (9%) (REN21, 2020). For the same years, Whiteman et al. (2020) state an increase of 48 GW (10%) to a total of 514 GW in 2017 and an increase of an additional 49 GW (9.6%) to a total of 564 GW in 2018. While the actual numbers in both reports differ slightly due to differences in their data basis, the data shows general trends common to both reports:

- The amount of produced wind energy shows an almost constant yearly linear increase with a mean value of approximately 49 GW (Whiteman et al., 2020) or 47 GW (REN21, 2020). The yearly increase varies strongly but shows a slight growth of approximately 4%–5% on average over the years.
- Offshore wind energy shows an average yearly increase of 2.16 GW (REN21, 2020) or 2.81 GW (Whiteman et al., 2020) taken as mean over the whole data sets. While these numbers are comparably small to the total number of installed wind energy capacity, offshore wind energy shows a much higher growth rate. Thereby, the increase in installed capacity grows by an average of 19% per year, according to both data sets.

Figure 2.2 shows the development of the percentage of produced renewable energy of Germany's total energy production (Fig. 2.2a) and the total amount of energy produced by onshore and offshore wind farms in Gigawatt per hour (Fig. 2.2b). This figure shows data provided by Germany's Federal Ministry for Economic Affairs and Energy (BMWi) in their regular publication of time series (Arbeitsgruppe Erneuerbare Energien, 2020).

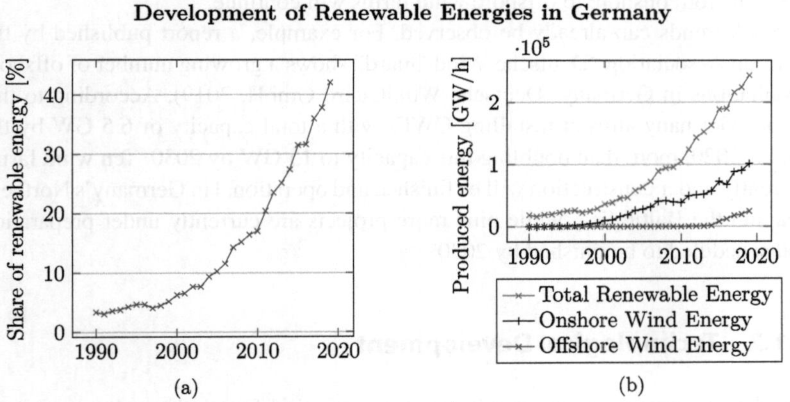

Fig. 2.2 (a) Share of renewable energy and (b) production of renewable energy, onshore, and offshore wind energy in Germany from 1990 to 2019. (Data from (Arbeitsgruppe Erneuerbare Energien, 2020))

The graphs show a rapid increase in the total production of renewable energy and a similar increase in the share of renewable energy in Germany's total energy production. Moreover, the graphs show that wind energy contributed approximately

52% of Germany's renewable energy production in 2019. Even though the first OWFs went operational 20 years after the time series started (2009), they already contributed approximately 20% of the produced wind energy in 2019.

The increasing focus on offshore wind energy results from several facts. On the one hand, OWFs produce more energy than their onshore counterparts due to higher wind speeds and -availability at sea (Breton and Moe, 2009; Sun et al., 2012). On the other hand, the open sea offers more space to install wind farms. The large number of existing onshore wind farms results in a decreasing availability of areas with high wind speed and availability, required to render the wind farm viable. Additionally, those areas become increasingly hard to access for new wind farm projects. Finally, onshore wind farms face several additional limiting factors, e.g., animal welfare regulations or local residents. For example, several countries state minimum distances between onshore wind turbines and settlements. In Germany, this distance currently ranges from 400 meters to 1.100 meters for larger settlements and from 300m to 1.000m from single houses, depending on the county (Fachagentur Windenergie an Land e.V., 2020). In 2020, government officials discussed introducing general legislation that would result in a minimum distance of one kilometer even between small settlements and the next wind turbine. In consequence, it can be expected that the shift from onshore to offshore wind farms will continue.

Such trends can already be observed. For example, a report published by the German Association "Deutsche Wind Guard" shows a growing number of offshore installations in Germany (Deutsche WindGuard GmbH, 2019). According to this report, Germany aims at installing OWFs with a total capacity of 6.5 GW by the end of 2020, more than doubling this capacity to 15 GW by 2030. Ten wind farms currently under construction will be finished and operational in Germany's Northern Sea and the Baltic Sea, while nine more projects are currently under preparation and scheduled to be finished by 2030.

2.1.2 Technological Development

Besides the increase in the number and size of wind farms, the technology itself has developed over the last years. In particular, wind turbines have become larger, heavier, and more capable. Figure 2.3 shows the technological development of onshore wind turbines. The figure depicts the average hub height and diameter of onshore wind turbines installed in the U.S. between 2006 and 2018. The numbers originate from a report published by the U.S. Department of Energy (Wiser and Bolinger, 2018). While the numbers show only a very slight increase in the average height of onshore turbines, the average rotor diameter increases by an average of 3.2 meters per year.

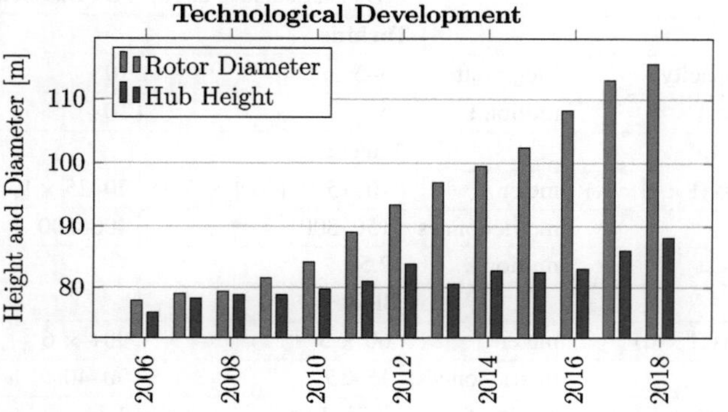

Fig. 2.3 Technological development of wind turbines in the U.S. for onshore projects. (Data from (Wiser and Bolinger, 2018))

The consulting Agency BVG Associates from the U.K. published two reports called *A Guide to an Offshore Wind Farm*, once in 2010 and an updated version in 2019. These reports provide several technical characteristics of offshore wind farms and turbines, which were common at the time of these reports. While only providing two data points, these reports demonstrate the technological progress made by OWTs. The values in Table 2.1 show the same trends of increasing dimensions and weights. Over the last decade, the average length of a rotor blade has increased by more than 50%. While the towers' height only slightly increased for onshore wind turbines from 78 to 88 meters, offshore turbines increased their height from approximately 80 to over 110 meters. The weight of most components nearly doubled for offshore turbines.

2.1.3 Summary

In conclusion, this section shows that wind energy gains increasing importance by contributing to renewable energy targets of countries worldwide. In Germany's example, this section shows that the number and size of planned OWFs increases in the near future. Moreover, components required to construct the turbines become increasingly larger and heavier. Studies show that these trends might result in bot-

Table 2.1 Common technical characteristics of offshore turbines and main components according to reports by BVG Associates from 2010 and 2019

Characteristic	Unit	2010 (BVGassociates, 2010)	2019 (BVGassociates, 2019)
		Turbine	
Default Capacity	Megawatt	3–5	10
Average Cost	million £	6	10
		Nacelle	
Dimensions (l × h × w)	meter	10–15 × 4 × 4	20–25 × 9–12 × 7–9
Mass	metric tonnes	150–300	400–500
Average Cost	million £	2.5	4
		Blades	
Dimensions (l × d)	meter	60 × 5	90+ × 6
Mass	metric tonnes	15–25	30–40
Average Cost	million £	0.75–1.05	1.3
		Tower	
Dimensions (h × d)	meter	80 × 4–6	110 × 5–8
Mass	metric tonnes	200–400	600
Average Cost	million £	1	0.7

tleneck situations regarding the availability of resources, i.e., installation vessels or port-side storage spaces (Beinke et al., 2020a; Oelker et al., 2020).

In the context of OWF, several authors highlight that logistics costs during installation make up a large proportion of the farm's life-cycle costs. Dewan et al. (2015) states a share of about 15%–20% and argues that these high costs result from the process' weather dependency. Lange et al. (2012) provide a similar conclusion and state that the logistics costs for the offshore installation highly depend on the current weather conditions, as vessels require high investments considering their charter rates. Muhabie et al. (2018) present a breakdown of the costs of an OWF, which they compiled from several sources. This breakdown attributes 5%–30% of the costs to the installation and transport. Despite the high costs for turbines given exemplary in Table 2.1, they state that turbines only contribute 30%–50% of the overall costs for OWFs. In contrast, turbines make up for 65%–75%, and logistics only for 0%–3%, for onshore wind farms. Consequently, this thesis focuses on the installation of offshore wind farms. Thus, the next section describes the installation process, possible influences, and related resources.

2.2 Installation of Offshore Wind Farms

Quandt et al. (2017) state that there exists no standardized process to install OWFs. Instead, there exist several concepts and strategies described in this section. Nevertheless, most of these concepts follow a similar basic structure and employ the same fundamental supply chain. The different concepts differ mainly in the way turbine components are handled and, thus, in the applied resources. The information provided in this section results from a literature review conducted with the keywords *installation, erection, offshore, wind farm*, which has been published in Rippel et al. (2019a). Additional information has been obtained from discussions with other researchers, who have experience in other research projects in the wind industry, and from additional literature not available when preparing that article. This review also provides the foundations for discussing existing approaches for the planning and scheduling of OWFs described later-on in Sect. 4.2.2.

2.2.1 Installation Process

Commonly, literature and practice separate the installation process into three phases, as, e.g., described in Vis and Ursavas (2016): The first phase includes the installation of founding structures and the connection of these structures to the energy grid. The second phase focuses on the installation of the so-called top-structures. Finally, the third stage subsumes the commissioning turbines:

Installation of foundations: This phase comprises two main activities. The construction of founding structures and the laying of underwater cables to connect the foundations to the next network hub. The installation of founding structures requires at least an installation vessel for the transport and the mounting of the structures. Depending on the type of founding structure, these vessels require different tools, e.g., hydro-hammers. Moreover, auxiliary vessels, e.g., submarines or vessels equipped to dampen the hydro-hammer's waves, join the installation vessel for these operations. Besides, this phase employs specialized cable vessels to attach and lay the underwater cables. Both activities rely on calm weather conditions, especially on low wave heights.

Installation of top-structures: This phase subsumes the installation of turbines, each consisting of a tower, one nacelle, three blades, and the installation of the hub and internal technical installations. As before, this phase mainly employs installation vessels, which usually just require large cranes, allowing them to work in heights of over 100 meters. Consequently, it also relies on calm weather conditions, particularly on low wind speeds, to safely conduct crane operations.

Commissioning: This phase subsumes the final testing and ramp-up of the turbines. This phase mainly relies on highly trained and certified personnel to perform the testing. In contrast to the previous phases, the commissioning does not rely on calm weather conditions but requires higher wind speeds for the tests.

In practice, companies often perform these phases sequentially (Beinke, 2018). The weather dependency of all installation operations results in a limited time per year, a so-called season where companies can perform the required operations. This season commonly ranges from April to approximately October, depending on the weather conditions (Rippel et al., 2019b). Consequently, companies often perform the first and the second phase in successive years, i.e., in the first year, they install the foundations, and in the second year, they construct the top-structures. This separation can also make sense from an economic point of view. Both phases rely on installation vessels but require different setups regarding the applied tools. Changing tools on these vessels takes much time, resulting in high setup times and costs, plus additional costs for chartering the required tools in parallel. In contrast, the third phase does not rely on these vessels or specific tools. In consequence, this phase can take place in parallel to the installation of top-structures. Despite this possibility, the ramp-up and testing rely on high wind speeds, which generally interfere with installation operations, often resulting in a timely separation of these phases.

Particularly the first two phases require components for the installation. The first phase mainly requires foundations and cables, while the second phase needs towers, nacelles, and blades. Figure 2.4 shows a wind turbine and marks the respective components.

The installation of the single components during the first and second phase progresses sequentially from bottom to top. Thus, companies first need to install the foundations before they can install the cables. During the installation of top-structures, they first need to construct the tower before placing the nacelle, hub, and finally blades.

Fig. 2.4 Offshore wind turbine and its components. (Photo: © Photocreo Bednarek / Adobe Stock Standard License)

2.2.2 Supply Chain

Due to the components' large dimensions and weight, specific production ports usually supply the components required for the construction. These ports denote production facilities with a connected port for the handling and storage of components. In consequence, the first and second phases rely on a geographically distributed supply network consisting of several production-ports, a base port, and the actual installation site. Figure 2.5 depicts the supply chain for the installation of top-structures following the conventional installation concept. Please refer to Sect. 2.2.3 for more details on different installation concepts.

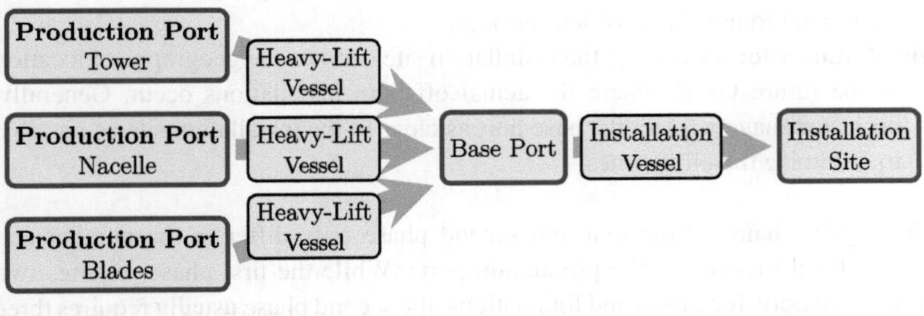

Fig. 2.5 Supply Chain and vessels for the conventional installation concept during the second phase. (Adapted from Rippel et al. (2019b))

The supply chain consists of three tiers: Production ports, the base port, and the installation site. The base port provides different services and acts as a decoupling point between production and installation. Consequently, this decoupling reduces the risk of disruptions in the production or transport impacting the installation. Vice versa, it also protects these parts of the supply chain from disruptions during the installation.

Production port: In general, each production site uses a nearby production port to ship a single type of component, e.g., founding structures, towers, or nacelles. These ports provide extensive storage capacities to buffer finished components, as production usually takes longer than the actual installation. Thus, they serve as decoupling points between the production and the following processes, i.e., transport and installation. Generally, companies can also decide to transport components onshore. Nevertheless, this thesis assumes offshore transports due to the components' large dimensions and the associated high logistic efforts to organize onshore transportation.

Base port: The base port acts as the central logistics hub and provides different services, e.g., for the maintenance of vessels or storage capacities for components. As production ports can be geographically far away from the actual installation site, the base port usually acts as centralized storage and buffer for components. Thus, most of the installation concepts assume that heavy-lift vessels transport components from the decentralized production ports to the base port, where installation vessels pick up the components before proceeding to the installation site. Some of the installation concepts also assume that different types of preassemblies take place within the base port. Project managers need to charter specific resources like heavy-duty storage spaces, loading bays, or other handling equipment to use the provided services.

Installation site: Generally, the installation site denotes the geographical location of the future OWF, where the actual offshore installations occur. Generally, project managers select the base port as close to the installation site as possible to minimize traveling times.

The supply chain for the first and second phase only differ in the number and geographical location of the production ports. While the first phase requires two production ports for cables and foundations, the second phase usually requires three ports for towers, nacelles, and blades. In some cases, these ports can be the same if the respective production sites are located close to the same port. In contrast to the supply chain described above, the third phase does not require additional components. While specialized equipment could be required for the testing and ramp-up, companies can obtain, store, and transport this equipment without the need for specialized resources.

2.2.3 Installation Concepts

The last section already mentioned the existence of several different installation concepts. These concepts describe alternative installation processes. Their differences range from a minor shift of assembly steps from the installation site to the base port to experimental concepts like floating substructures. In practice and literature, the conventional installation concept still holds a dominant role and also forms the baseline for the other concepts found in the literature:

Conventional installation concept: The conventional installation concept follows the supply chain described in the last subsection: After manufacturing, production ports store the components. Heavy-lift vessels pick up the components and transport them to the base port for buffering. Before the actual installation starts, the base port establishes a sufficiently high initial inventory level to ensure the availability of components for the installation phase. During this phase, installation vessels travel between the base port and the installation site. Thereby, they collect a number of component sets, depending on their storage capacity, and proceed to the installation site. At the site, the installation vessels install the components successively. After finishing a turbine, the installation vessels either return to the base port or reposition within the wind farm and begin constructing another turbine. This decision depends on the remaining component sets loaded on the vessel and the weather conditions.

This concept provides several advantages. As a central logistics hub and supply chain decoupling point, the base port provides flexibility and reduces disruptions along the supply chain. By providing an initial stock, it reduces the risk of component shortages. Moreover, it reduces the amount of heavy-duty storage required at each production port, equalizing capacity requirements across the supply chain. If current weather disruptions impede the installation process, the base port can mitigate some of the effects by providing an additional buffer for components. As a drawback, the base port adds additional costs in terms of chartering spaces and equipment. Moreover, installation vessels, which constitute a significant cost factor during the installation, perform non-value-adding tasks, i.e., transporting components between the base port and the installation site.

Feeder ship concepts: Feeder ship concepts try to increase the cost-efficiency of installation vessels by mitigating trips between the base port and the installation site in the so-called base port feeder concepts (Ait Alla et al., 2017). This concept assumes that the installation vessel remains at the installation site and that specially equipped feeder vessels deliver the components from the base port to the vessel. The offshore feeder concept even goes one step further. It applies feeder vessels to deliver the components directly from the production ports to the installation vessel, mitigating the need for a base port altogether (Oelker et al., 2018).

The advantage of this approach results from the extended time that installation vessels can spend with value-adding tasks. Moreover, the removal of the base port can lead to massive cost reductions. Nevertheless, the removal of the base port increases the risk of supply disruptions. Feeder vessels need to realize a just-in-time supply of the installation vessel. Moreover, these feeder vessels require capabilities to perform offshore transshipment operations. While the installation vessel supports such operations, feeder vessels still need additional technology to stabilize themselves for transshipment. In consequence, their prices can be assumed to be higher than standard heavy-lift vessels. Moreover, transshipment operations require comparably calm weather conditions, which reduces the efficiency of this approach.

Pre-assembly concept: Pre-Assembly concepts, e.g., proposes by Lange et al. (2012), Ahn et al. (2017), or Vis and Ursavas (2016), assume that companies already perform some assembly steps at the base port instead of at the construction site. The preassembly of tower segments to the complete tower constitutes a preassembly often performed in practice. The authors of the cited articles also propose and evaluate different other types of preassemblies, e.g., rotor star or bunny-ear assemblies, which denote preassemblies of two or three blades and the nacelle.

On the one hand, preassembly strategies reduce the risk of weather-related disruptions, as installation vessels need to perform fewer offshore operations per turbine. On the other hand, preassemblies require additional storage spaces and more careful handling due to their increased dimensions and weight. Vis and Ursavas (2016) state that such strategies can be beneficial for larger OWFs but requires careful capacity planning.

Floating structure concept: Finally, floating structure concepts rely on new technologies for the turbines founding structures. These floating structures allow turbines to float on the water surface with only a loose coupling to the seafloor (Arshad and O'Kelly, 2013). Such floating turbines allow for a complete preassembly in the base port. Afterward, tug boats drag the finished turbine to the OWF, where specialized vessels perform the anchoring and final grid connection.

The floating structure concept can be considered a different technology, associated with a different installation process than the other concepts. Floating wind turbines thereby provide distinct advantages, mainly that they can anchor further out in the sea. While conventional founding structures need a fixed coupling to the seafloor, these floaters only require a loose coupling. Unfortunately, most of these floaters remain in development or a prototypical state. Moreover, up to now, these turbines provide less capacity due to limitations in their height and rotor star diameter.

The concepts presented in this subsection all provide different advantages and disadvantages. While the literature has studied all concepts, practice mainly applies the conventional installation concept. Even if the base port and the installation vessel's trips incur additional costs, the advantages of this concept's high flexibility and the additional decoupling point within the supply chain mitigate the risk of unplanned delays and costs.

2.2.4 Classes of Vessels

The installation process employs a variety of vessels throughout its different phases. Thereby, these vessels differ in terms of their operations, equipment, and characteristics like speeds, capacities, and costs (Quandt et al., 2017). The actual type of vessel depends on the selected installation concept.

Heavy-lift vessels: These vessels generally take over the transport of components from the production ports to the base port. They provide high storage capacities and decent traveling speeds. Heavy-lift vessels possess no specialized equipment like cranes, as the port's cranes or loading bridges usually take over the actual loading and unloading operations. According to Meyer (2014), these vessels have comparably low charter rates of about 10,000€ to 20,000€ a day.

Feeder vessels: Feeder vessels constitute a sub-class of heavy-lift vessels applied in feeder-based installation concepts. As these concepts rely on offshore transshipment for the components, these vessels require technology to precisely position and stabilize themselves close to the installation vessels during these operations.

Crew vessels and helicopters: While the first and second installation phases generally employ installation vessels to transport crews, material, and components, the commissioning phase generally relies on standard crew vessels to bring teams to the installation site. In urgent cases, especially if a turbine requires maintenance, companies use helicopters to quickly transport specialists to the installation site.

Cable vessels: These vessels provide capabilities to transport and lay underwater cables over long distances. In consequence, these vessels are used during the first installation phase to connect the foundations to the energy grid. Figure 2.6a shows such a cable vessel.

Installation vessels: Commonly, so-called jack-up vessels or barges act as installation vessels. Figure 2.6b shows an example of such barges. These vessels provide the unique capability to use retractile pillars to mount themselves into the sea bed and raising themselves above the water surface. This so-called jack-up operation effectively turns the vessels into a stationary offshore platform, mitigating the influence of high waves on the installation operations. In contrast to so-called jack-up platforms, vessels are self-propelled. While companies might also use jack-up platforms, a regular relocation would require additional tug boats on stand by. Moreover, jack-up vessels are usually equipped with a large crane to allow installation operations in 100+ meters height. These vessels can be equipped with additional tools, e.g., for the installation of foundations and provide extensive storage and transportation capacities. Charter rates for jack-up vessels in the literature range from 70,000€–145,000€ (Meyer, 2014) to 108,680€–327,750€ per day (Dalgic et al., 2013), when applying a currency conversion rate for pounds to euro of 1:1.14 on April 10th, 2020. Consequently, these barges constitute a major cost factor for the installation of OWFs.

(a) (b)

Fig. 2.6 Vessels used during the installation of OWFs: (a) Cable vessel (© Van Oord) (b) Installation Vessel. (© Mykolayiv/Adobe Stock Standard License)

The description of employed vessels shows that jack-up vessels take on a key role during the installation as they perform most value-adding offshore operations. The high daily charter rate of these vessels requires careful planning and scheduling of operations. Thereby, contractual charter rates often differ by up to 30% depending on the vessel's state, i.e., if the vessel is currently offshore or in port (standby) (Lange et al., 2012; Rippel et al., 2019d). Moreover, project managers need to reserve some of these vessels up to two years in advance (Kerkhove and Vanhoucke, 2017).

2.2.5 Dynamic Influences on the Installation Process

Besides the mentioned weather influences, several different aspects impact the overall efficiency of the described supply chain. Schweizer et al. (2011) and Beinke (2018) describe different dynamic influences and disturbances affecting the supply chain, summarized in the following:

Weather: Weather poses the primary influence on the overall supply chain performance. As noted, installation operations are highly weather dependent, while jack-up vessels contribute a significant cost factor during the installation process. Consequently, the next subsection provides a broader overview of these influences and the operations' restrictions.

Processing times: Even though the last decade has shown an increase in installed OWFs, the technology itself still advances quickly in terms of dimensions, weight, and complexity. In consequence, processing times along the supply chain contribute high degrees of uncertainty. As installation projects differ in their applied technology, each project shows distinct characteristics in processing times and learning curves, impeding precise production and installation planning.

Transportation delays: The large dimensions and weight of the turbine components require specialized transportation vessels and trucks. Companies need to apply for these transports permissions, particularly if a project requires onshore transport to the production ports. Consequently, deviations and delays during the transport can induce disruptions to the original plan, requiring suitable safety stocks at each port.

Resource breakdowns/damaged components: While resource breakdowns or damage to components rarely occur in this supply chain, each incident results in long delays and disruptions due to the high specialization of components and resources.

Missing components/production delays: Comparable to damaged components, missing components can cause severe delays in the installation process. Such situations might occur due to mistakes in the production or transportation planning, order mismatch, or mix-ups during the transport.

Workforce scheduling and qualification: Offshore operations require specially trained and certified personnel. This risk mainly impacts the commissioning stage as it requires highly trained technicians. Nevertheless, possible sickness may introduce additional uncertainty to the process planning and execution.

This summary lists several possible sources of disruptions and uncertainties that can impact the installation process. Most of these sources are common to all supply chains, e.g., breakdowns, delays, or varying processing times. Nevertheless, if such an incident occurs, it can be assumed that it impacts the offshore supply chain more severely due to the size, weight, and complexity of the components and resources. Companies cannot simply replace a loading bay or jack-up vessel that needs unscheduled maintenance. The same accounts for components. In both cases, rework/maintenance or replacements can take days to months.

2.2.6 Weather Restrictions

Besides supply chain risks, which are common to most supply chains, the last subsection highlights weather dynamics as the primary source of uncertainty to the process planning and scheduling. This subsection presents an in-depth discussion of weather restrictions applicable to the installation process.

Generally, most offshore operations require stable weather conditions to safely lift and install components in heights of approximately 100 meters. In this context, literature and practice apply strict restrictions in terms of maximum wind speeds and wave heights for each operation to ensure safe handling of the components. For example, during the nacelle's installation, the vessel's crane lifts the nacelle on top of the already installed tower, where technical personnel positions and fixes the nacelle. High wind speeds or an unsteady vessel can set the nacelle and the tower in motion, posing a risk to the personnel, the nacelle, and the already installed tower. Consequently, if the current conditions exceed these limits, the respective operation cannot start or proceed. Moreover, operators need to abort the operations and restart them later when the conditions become safe again if the operation has already started. All of the methods for the scheduling or simulation of the installation process found in the literature include weather restrictions in one or another way (c.f. Sect. 4.2.2). Table 2.2 lists the weather restrictions found throughout the literature.

Table 2.2 Operations and weather restrictions in the literature (Rippel et al., 2019a)

Operation	Wind [m/s]	Wave height [m]	Ref.
Sailing	21	3	1,6
	17	4	7
	16	–	2
	12	4.8	3,5
Installation Towers / Nacelles	17	2	7
	12	2.5	1,4,6
	10	–	2
	6.5	2.5	3,5
Installation Blades	17	2	7
	10	2.5	1,2,4,6
	6.5	2.5	3
	5	2	5
Installation Foundation	18	3	4
	17	2	7
	11	3.5	3,5
Installation Piles	17	2	7
	11	3.5	3
Installation Cables	11	3.5	3
Un/Loading Towers / Nacelles in port	17	10	7
	14	5	1,4,6
	10	–	2
Un/Loading Blades in port	17	10	7
	12	5	1,4,6
	10	–	2
Un/Loading Foundations in port	18	5	4
	17	10	7
Seafastening	14	2	1,6
	14	1.4	7

(continued)

Table 2.2 (continued)

Operation	Wind [m/s]	Wave height [m]	Ref.
Transship Towers/Nacelles	14	2	1,6
Transship Blades	12	2	1,6
All Operations for floating structures	10	4	7

1 (Oelker et al., 2018)
2 (Vis and Ursavas, 2016)
3 (Ait Alla et al., 2013)
4 (Beinke et al., 2017)
5 (Quandt et al., 2017)
6 (Ait Alla et al., 2017)
7 (Dewan et al., 2015)

The table shows that authors in the literature apply differently strict limitations to each operation. Moreover, the table shows that the wave height does not impact operations within the base port (loading/unloading). Nevertheless, assumptions for the wind speed differ between comparably high values of 17 m/s and restrictive values of 10 m/s. Within the base port, companies can apply different technologies, e.g., fixed loading bridges, to partially mitigate high wind speeds. Nevertheless, loading/unloading still requires some sort of crane operation to lift components weighing several tons onto or off the vessels. Wave height restrictions for offshore operations only differ slightly within the literature. In contrast, wind speed limitations cover an extensive range depending on the source. Thereby, the limits provided by Ait Alla et al. (2017); Beinke et al. (2017); Oelker et al. (2018) rely on experience and discussions with companies involved in offshore installation projects and will thus be used as baseline assumptions throughout this thesis.

2.3 Planning Tasks

The process description given in the last section refers to several planning tasks required to install an OWF successfully. This section presents a decomposition of the overall planning problem into smaller tasks with varying planning horizons, as given in Rippel et al. (2019a). Figure 2.7 provides an overview of these tasks.

Tasks on a lower level (shorter horizon) generally depend on information and decisions from higher levels. In contrast, higher-level decisions often rely on reasonable estimates or predictions resulting from lower-level planning tasks. For example,

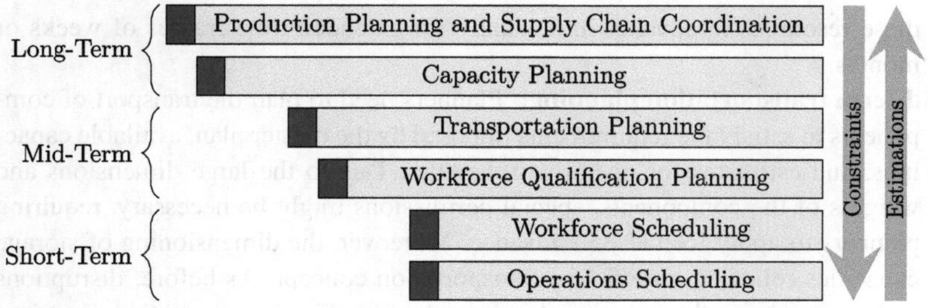

Fig. 2.7 Decomposition of planning tasks

to efficiently dimension the base port's storage capacity, planners need information on how many turbines could be built and delivered in a certain amount of time. The other way around, an estimation of these amounts requires an estimate of the base port capacity as, e.g., installations require sufficient components to proceed quickly.

Long-term production planning and supply chain coordination: This task includes determining a suitable supply network, i.e., choosing base and production ports depending on the installation site's location. Moreover, project planners need to establish a master production plan. Particularly feeder concepts, which rely on just-in-time deliveries, need a precise production plan. On the one hand, this master plan requires sufficient buffers in terms of times, materials, and resources to ensure feasibility. On the other hand, it needs to be as precise as possible to reduce extensive charter rates for heavy-duty storage areas and vessels. Within the literature, most authors assume that the base port provides sufficient components and that the supply network is fixed. Thus, literature generally excludes this planning task.

Long-term capacity planning: This task aims to determine suitable amounts of resources. Resources include heavy-duty storage at the ports, vessels, handling equipment, or tools. Dynamics and disruptions on lower-level planning tasks highly influence the capacity planning. For example, if the weather prevents installations as planned, additional storage might be required. In contrast, if the weather allows for a faster installation as originally predicted, excess capacity induces unnecessary costs to the project. Several methods can handle limited storage or transportation capacities or focus on the evaluation of varying fleet sizes within the literature. Most of these approaches only account for costs for the resources' active times, neglecting that in practice, planners need to charter these resources in advance and usually for extended time frames of weeks or months.

Mid-term transportation planning: Planners need to plan the transport of components to satisfy the requirements imposed by the master plan, available capacities, and estimates for operative schedules. Due to the large dimensions and weights of the components, special permissions might be necessary, requiring planners to apply for these in advance. Moreover, the dimensioning of storage capacities relies on an efficient transportation concept. As before, disruptions and dynamics on the operative level can highly affect transportation planning. For example, deteriorating weather conditions might impede planned transports on waterways, resulting in component shortages. Only very few articles exist regarding the supply of components between production ports and the base port in the current literature.

Mid-term qualification planning: Particularly, the commissioning stage requires highly qualified personnel. Nevertheless, several companies require all offshore personnel to undergo certification before working on the turbines. Consequently, this task aims to ensure that the installation project can access sufficient personnel with the required certification. Similar to the capacity planning, this task is highly susceptible to disruptions and dynamics on the operative level.

Short-term operations and workforce scheduling: Project planners can only schedule actual offshore operations and the related personnel requirements for a short time-frame due to the strong influence of dynamic weather conditions. In practice, planners, port operatives, and the vessel's captains rely on weather forecasts and their experience. According to discussions with other researchers in this field, this scheduling generally takes place ad-hoc, i.e., decisions are made daily, sometimes hourly, without any long-term planning. These short time frames and the uncertainty involved with weather forecasts induce high uncertainty to higher-level planning tasks. In the literature, most articles focusing on the scheduling or simulation of offshore operations include weather dynamics in one way or another. Unfortunately, none of the articles found during the literature review includes forecast uncertainties. Instead, all of these approaches assume that weather conditions are either known a-priori or simply "react" to changing conditions by rescheduling.

In conclusion, higher-level tasks require estimates covering a broad time frame, spanning months or even longer periods. Their resulting decisions stay valid for the full duration of the installation project. While it is theoretically possible to change the supply chain's layout or charter additional capacity, the involved processes can take a long time and may even exceed the planned project duration. Thus, these tasks only provide minimum flexibility once planners make these decisions. Mid-level tasks provide a more considerable degree of flexibility. Nevertheless, due to requiring certification and permissions, decisions made on this level still require careful planning and cannot be changed instantly. In contrast, low-level decisions refer to scheduling tasks. These schedules highly depend on the weather and, thus, will frequently change. These changes induce high uncertainty for higher-level tasks, which rely on these schedules to enable efficient planning.

2.4 Research Focus and Requirements

This section aims to outline this thesis's research focus and the requirements arising from the presented process description and decomposition. Therefore, it shortly summarizes each subsection and describes its implications on the research focus.

2.4.1 Research Focus

The decomposition of the planning problem in smaller planning tasks shows that offshore installation projects need precise planning on different time horizons. Thereby, higher-level tasks rely on estimations for lower-level tasks, and vice-versa provide constraints for these lower-levels. Moreover, weather conditions during the project execution affect all planning tasks directly or indirectly, introducing a high level of uncertainty and dynamics. Consequently, the proposed methodology mainly focuses on the *scheduling of offshore operations*. Nevertheless, later parts of this thesis will evaluate the applicability of offline schedules for higher-level planning tasks, e.g., the capacity and transport planning.

Literature and practice separate the installation of OWFs into three main stages: the installation of foundations, the construction of the top-structures, and the commissioning. While these phases can be performed concurrently, the use of similar resources, i.e., jack-up vessels and the dependency of later stages on the previous ones, justify a successive execution of the stages. This thesis focuses on the *installation of top-structures*, as this stage comprises a greater variety of components. Nevertheless, the proposed methodology can be applied to installing foundations without limitations, as the same operations occur, but with different times and restrictions. An application to the commissioning stage is possible if reducing the problem to only scheduling the workforce. Nevertheless, comprehensive scheduling during this stage requires a different model formulation to combine workforce and vessels into the operation-scheduling.

In consequence, this thesis mainly focuses on the latter part of the supply chain. Generally, the supply network for the construction of top-structures consists of the installation site, the base port, and three production ports. The scheduling mainly focuses on the *base port and the installation site*. Later parts of this thesis will include the production ports when evaluating the use of generated offline schedules for transportation and capacity planning.

This section presented several installation concepts found in the literature. This thesis focuses on the *conventional installation concept*, as it is the most applied in theory and practice. Nevertheless, the proposed methodology can easily handle preassembly concepts by modifying process chains and restrictions. The proposed methodology does not support feeder concepts. In contrast, the methodology could be used to schedule floater-based installations by reducing the process chain to a single installation operation and by adjusting the weather restrictions.

In terms of dynamic influences, this thesis mainly focuses on handling *weather dynamics* and their impact on processing times. On the one hand, these dynamics require specialized approaches to handle and are quite unique to the offshore area. Consequently, there are not many approaches to deal with these dynamics. On the other hand, the remaining dynamics described in the previous subsection are common to mostly all supply chains. While these dynamics might introduce severe disruptions into the installation process, they rarely occur in this context.

2.4.2 Design Features and Requirements

Due to the significant uncertainties and high costs involved in offshore installation projects, classical robust approaches, which generally plan for the worst-case, seem unsuitable for this planning problem. Consequently, this thesis aims to propose and evaluate a methodology, which can be used as part of a *decision support system* for the process planning and scheduling of offshore installation projects. Table 2.3 lists several design features and their resulting functional requirements imposed on this DSS.

The underlying methodology needs to handle *weather forecasts and uncertainties*, allowing precise planning while accounting for the associated uncertainties. Moreover, it can be assumed that project planners and port officials will use such a tool to render planning or scheduling decisions. As most methods in the optimization and simulation area require experts to set up and modify, this thesis aims to evaluate to which degree the DSS's underlying models can be customized via *model transformations*. Thus, project managers could use a content driven interface, describing the actual project, while model transformations can be applied to generate the simu-

Table 2.3 Design features and resulting requirements for the developed methodology

Feature/Requirement		Description
	Decision support system	
R1	Solution quality	The algorithm should deliver as good as possible plans. While the online scheduling might be limited by its real-time constraint, the offline scheduling should aim for optimal solutions.
R2	Real-time capability	Especially online scheduling requires low computational times. Additionally, the solutions must provide a long validity, i.e., the number of recourse actions should be minimal.
R3	Planning tasks with diff. horizons	Results of the scheduling should be reusable for higher-level planning tasks.
	Targeted users: Project planners	
R4	Project focused interface	Users should need minimal knowledge about the underlying algorithms and models.
	Online scheduling	
R5	Weather forecasts	The methodology should include methods to preprocess the stochasticity of forecasts to obtain good results and reduce the number of recourse actions.
R6	Reactivity	Online scheduling requires an incremental approach to handle weather dynamics.
	Offline Scheduling	
R7	Historical weather data	The methodology should include methods to preprocess and aggregate historical data.
R8	Long-term schedule	Offline scheduling requires methods to generate long-term plans within reasonable time frames.

lation/optimization models. Finally, the decomposition shows that the scheduling of operations has a predominant influence on all other tasks. On the one hand, disruptions during the project execution can invalidate all higher-level plans. On the other hand, most of the other tasks rely on a precise estimate of the installation progress, e.g., to plan capacities, the production, or transports accordingly. Consequently, the DSS focuses on the scheduling of offshore operations. Thereby, this thesis distinguishes between two options: online scheduling and offline scheduling. *Online scheduling* denotes process accompanying scheduling, where a scheduling algorithm uses currently available information to optimize a plan for the (near) future. In this thesis, the term *offline scheduling* denotes the calculation of the estimations mentioned above. The scheduling algorithm needs to be capable of applying aggregate historical weather information to obtain estimated schedules for higher-level planning tasks.

The DSSs should satisfy several functional requirements to provide users with added values. It needs to provide good, in the best case, optimal solutions within a feasible time frame. Thereby, a feasible time frame in this context can cover one or more hours, as most operations in the offshore context require several hours to complete. Finally, while the methodology mainly focuses on the scheduling, the results should be applicable to other planning tasks as well.

The targeted user base of project planners, port officials, and maybe the vessel's captains imposes the functional requirement that the usage of the methodology, or, more precisely, the resulting DSS, should only require minimal knowledge about the underlying simulation/optimization models. In contrast, these users require an interface, which allows them to model and input all relevant information about the project.

The methodology's design will include methods for online and offline scheduling. Both approaches pose different functional requirements. Online scheduling requires an iterative (reactive) planning approach, capable of processing weather forecasts due to their inherent stochasticity. In contrast, offline scheduling aims to deliver an estimate of probable schedule using historical information. Therefore, it requires methods to provide a long-term schedule and aggregate and process the stochasticity of larger amounts of historical data.

2.4.3 Proposed Methods

This subsection proposes several methods as part of the overall methodology to satisfy the given requirements and design features. Based on this selection, the next chapter introduces these methods in more detail, presents the state of the art on these methods' applications in related scenarios, and justifies their use. Table 2.4 provides an overview of the proposed methods and approaches. The table shows three main classes of methods: Optimization, more precisely the use of Mixed-Integer Linear Programs (MILPs), incremental approaches, more precisely the application of a Model Predictive Control (MPS) scheme, and the application of model transformation and meta-modeling to provide a suitable user interface. Besides, statistical approaches will be used to handle uncertainties induced by weather data and forecasts.

Table 2.4 Proposed methods and approaches for each requirement

Feature/Requirement		Proposed Methods
Decision support system		
R1	Solution quality	Optimization (MILP)
R2	Real-time capability	Time restrictions on the optimizer
R3	Planning tasks with diff. horizons	Meta-modeling
Targeted users: Project planners		
R4	Project focused interface	Model transformations and code generation
Online scheduling		
R5	Weather forecasts	Statistics to handle uncertainties
R6	Reactivity	Incremental approaches (MPC)
Offline scheduling		
R7	Historical weather data	Statistics for data aggregation
R8	Long-term schedule	Optimization (Scheduling/MILP)

3 Methodological Basics

As noted in the last chapter's conclusion, this thesis applies methods from the areas of control theory and scheduling, particularly MPC and MILP, to propose a methodology that can act as DSS for the installation of offshore wind farms. Consequently, the following sections provide a short introduction to these topics to justify selecting these particular methods. Section 3.1 first introduces different types of DSS and discusses their characteristics. Afterward, Section 3.2 provides an overview of different types and classes of scheduling algorithms. These algorithms provide the foundation for the later DSS. Section 3.3 presents an overview of rolling horizon planning with a focus on the MPC scheme. This thesis applies this scheme's predictive-reactive characteristics to include methods for the handling of weather uncertainties. Finally, Section 3.4 introduces the notion of model transformations and meta-modeling used to decouple the DSS's underlying methodology from the user interface.

3.1 Decision Support Systems

Decision support systems have a long history and have gradually developed further since the 1970s (Shim et al., 2002). In the literature, several authors state functionalities and characteristics of DSSs, whereby Dulcic et al. (2012) summarizes characteristics in three primary functionalities:

Specificity: DSSs focus on a specific decision-making process and task. Developers need to specify the aim and develop the underlying mechanics accordingly.

Interactive support: DSSs provide users with interactive help to render decisions but do not automate decision making.

© The Author(s), under exclusive license to Springer Fachmedien Wiesbaden GmbH, part of Springer Nature 2026
D. Rippel, *Online Decision Support for Offshore Wind Farm Installations*, https://doi.org/10.1007/978-3-658-49912-9_3

Adaptivity: DSSs can adapt to the users' needs, including an adaptation of the scenario and integration of current information and data.

By evaluating existing literature on the topic of DSSs, Power and Sharda (2007) define five basic types of DSSs: model-driven, communications-driven, data-driven, document-driven, and knowledge-driven DSSs. Each of these systems aims at a different kind of decision support. For example, communication-driven systems aim to support shared decision making between different stakeholders, while data-driven systems aim to analyze vast amounts of data, e.g., to provide business intelligence. The authors state that these DSSs provide non-technical personnel with the ability to perform "what if?" analyses using accordingly developed models. The authors highlight that such DSSs differ in two ways from classical computer support tools, further refining the main characteristics given above:

Accessibility: Model-driven DSSs need to provide an easy-to-use user interface for non-technical personnel, e.g., managers.

Repeated use: Model-driven DSSs allow a reapplication of the system to the same or similar tasks.

In a more specific context, Hazir (2015) states that in project management, DSSs mainly adopt the role of project monitoring systems. Thus, they aim to minimize the deviation of projects from a given plan. This task includes reporting the project's state, the comparison against planned progress, analysis of deviations, and, finally, the deduction of corrective measures. Thus, DSSs for project management implement two kinds of policies:

Monitoring policy: This policy clearly describes which values the system monitors in what intervals and how it obtains these values.

Intervention and control policy: The intervention and control policy defines a set of tools and recourse actions the DSS applies to correct deviations from the plan.

In conclusion, this thesis proposes a methodology suitable for developing a model-driven DSS using a combination of MILP-optimization and an MPC scheme to provide decision support for offshore installations. Therefore, it applies a model of the installation process to propose schedules or other results for higher-level planning tasks. It mainly focuses on scheduling operations, both online (operative DSS) and offline (strategic or tactical DSS).

3.2 Scheduling

The process description implies similarity to common scheduling problems, which assign a defined number of jobs, each consisting of a set of sequential operations to a fixed number of capable machines or, in this case, vessels. Most approaches in the literature apply a minimization of the makespan as their cost function, i.e., they minimize the time required to complete all jobs (Chaudhrya and Khanb, 2016).

Comparing this classical notion of scheduling problems to the installation of OWFs, vessels take over the role of machines, while assembly and movement operations take the place of manufacturing operations. A single job represents the installation of a single OWT, including its loading operations. The scheduling problem's structure resembles a so-called job-shop scheduling problem (JSSP), as there is no fixed sequence between machines/vessels as it would be typical for flow-shop problems. JSSPs belong to the class of NP-hard, or even NP-complete problems, if using three or more machines/vessels (Garey et al., 1976). Thus, an optimizer cannot guarantee to find an optimal solution within polynomial time or only requiring a polynomial amount of space in relation to the dimensions of the scenario. Consequently, literature proposes a multitude of solution algorithms for JSSPs, ranging from exact mathematical methods to meta-heuristic search algorithms. Zhang et al. (2019a) present a current review of different solution algorithms. Moreover, several authors provided a variety of classifications for JSSPs regarding their formulation and their characteristics. The following subsections present several sub-classes of JSSPs to categorize the problem at hand.

3.2.1 Classification of Job-Shop Scheduling Problems

The literature proposes a variety of classifications for JSSPs, depending on their characteristics and characteristics of their environment, e.g., Abdolrazzagh-Nezhad and Abdullah (2017); Herroelen and Leus (2007); Zhang et al. (2019a). Thereby, Zhang et al. (2019a) propose a classification according to four major types of formulations:

Basic JSSP: The basic JSSP constitutes the base-class of job-shop scheduling problems. It assumes that only a single, predefined machine can perform each operation. Consequently, this class describes a sequencing problem (order of operations per machine).

Flexible JSSP: Flexible JSSPs extend the basic formulation by assuming that a defined set of alternative machines can perform multiple operations. Consequently, the problem's complexity increases: In addition to the sequencing problem, these problems include an additional routing problem in deciding which machine should perform an operation (Chaudhrya and Khanb, 2016). This class further distinguishes between complete and partially flexible JSSPs. For complete problems, each resource can perform all operations. Partial problems assume that resources can perform overlapping sets of operations, even though some operations remain specific to only some resources.

Multi-Resource JSSP: Multi-Resource JSSPs extend Flexible JSSPs even further. In addition to machines, each operation requires other limited resources, e.g., casts.

Multi-Plant Multi-Resource JSSP: The most complex class of JSSPs additionally regards multiple plants or production facilities. Therefore, it needs to regard and plan additional transports of orders between those facilities, relying on limited transportation resources.

According to the above classification, the installation of OWFs considered in this thesis resembles a complete flexible JSSP as projects can employ several (alternative) installation vessels capable of performing all operations. Only the installation of cables and the commissioning of finished OWTs require different resources. Nevertheless, each operation requires a specified type of resource, and no overlaps exist between these sets, which would result in a partially flexible JSSP.

The DSS might require more complex classes of JSSP when considering the commissioning or multi-site projects. The commissioning might need an extension to a multi-resource JSSP to plan personnel and transportation resources simultaneously. In contrast, multi-site projects might not need an extension of the formulation. Each job (turbine) belongs to a specific installation site, and orders (components) would not need to pass from one site to another. Consequently, such projects only incur additional traveling times or operations for the installation vessels.

3.2.2 Formulations for Flexible Job-Shop Scheduling Problems

In the literature, authors propose a variety of formulations and models for flexible JSSPs. Nevertheless, these generally apply one of three basic formulation types. Demir and İşleyen (2013) describe these basic formulations and compare their advantages and disadvantages in terms of computational time and complexity against each other:

Time-Indexed formulation: These models generate schedules using discrete time instances, e.g., minutes, hours, or days. For each machine and time instance, these schedules denote which operation the resource currently processes. Compared to other formulations, this class requires a high number of decision variables, which is at least quadratic over the number of machines and time instances. Following Demir and İşleyen (2013), these formulations require an estimation of the planning horizon's length so that the optimizer can schedule all operations.

Sequence-Position formulations: Instead of time instances, these formulations use discrete slots referring to semantic concepts, e.g., to the maximum duration of machining operations or shifts. Schedules represent each machine as a sequence of slots, assigning operations to them. While closely resembling time-indexed formulations, these formulations reduce the number of decision variables by summarizing them into discrete slots of equal length. For example, assuming machining operations with durations of 2, 2.5, and 3 hours would result in slots of three hours. While reducing the number of variables, this representation often results in weaker solutions as some operations actually require less time than scheduled. Similar to time-indexed models, these formulations require approximating the number of slots required to schedule all operations. Basically, this class summarizes all indexed formulations focusing on higher-level concepts instead of discrete time units.

Precedence-based formulations: In contrast to the other types of formulations, precedence-based models rely on continuous time representations. The resulting schedules index their solutions over the number of operations, usually resulting in a minimal number of decision variables. Consequently, these schedules denote start or end times for each operation on the assigned machines. While this type of formulation does not need an estimation of the overall duration, the optimizer always needs to schedule all operations and jobs, no matter how long the desired planning horizon would be.

In addition to the type of formulation, the scheduled system's environment poses an essential factor for the choice of viable approaches. Several authors, e.g., Zhang et al. (2019a) or Wojakowski and Warzolek (2014), propose to differentiate between static and dynamic JSSPs. Static scenarios assume that all variables and conditions are known in advance. Contradictory, dynamic scenarios cover various dynamics like variable processing times or machine breakdowns, which generally induce some sort of uncertainty or stochasticity into the scheduling.

3.2.3 Scheduling under Uncertainty and Dynamic Scheduling

The process described in the previous section clearly shows that the installation of OWFs depends on various external dynamics and, thus, should be considered a dynamic JSSP. As before, the literature proposes several approaches and classifications for dynamics formulations. Several authors provide reviews of current methods for dynamic scheduling in production systems, e.g., Dias and Ierapetritou (2016); Ouelhadj and Petrovic (2009); Wojakowski and Warzolek (2014). These reviews provide similar classifications according to the classes reactive, preventive, and predictive:

Reactive and Continuous Approaches: Reactive approaches generally omit the inclusion of uncertainties into the initial planning. Consequently, these approaches apply so-called recourse actions in cases of plan failures, i.e., rescheduling. Therefore, some authors distinguish between approaches, which react to disturbances and those that perform continuous rescheduling. Generally, approaches of the reactive kind focus on reducing rescheduling times, e.g., by explicitly selecting affected operations or applying dispatching rules. Continuous approaches often use receding horizon techniques and combine these with state-space descriptions to avoid expensive optimizations.

Preventive (Robust and Stochastic) Approaches: Preventive approaches aim to provide schedules that incorporate the system's dynamics. Such plans generally focus on the worst possible case (robust plans) or provide alternative samples for probable scenarios using stochastic programming techniques. Consequently, stochastic approaches generally require long computational times as they compute schedules for several probable scenarios, possibly even for several time steps. Robust approaches omit these high computations costs by applying a worst-case assumption, often resulting in very conservative schedules.

Predictive-Reactive Approaches: These approaches try to estimate the extent of disturbances in the scheduling and, additionally, apply reactive recourse actions in case of unexpected disturbances. For example, Dias and Ierapetritou (2016) describe fuzzy-based approaches, translating uncertainties into a fuzzy description, trying to determine the most likely values for each variable. This reformulation reduces the models' complexity similar to robust schedules but omits the worst-case assumption. Ouelhadj and Petrovic (2009) provide an overview of approaches using bi-objective formulations, minimizing the makespan while reducing the impact of disturbances on the schedule.

Bakker et al. (2019) present a review of general techniques for scheduling under uncertainty. Thereby, they mainly focus on approaches to handle multi-periodic problems, i.e., problems where the uncertainty and the state of the scheduled system change over time. The review shows a similar list of common techniques as the classification provided above. According to the authors, most of the literature applies stochastic programming, robust optimization, online (continuous) optimizations, or combinations of these to cope with uncertainty. Next to an in-depth description of different methods and the historical evolution of each technique, the authors describe the results of a broad literature review for applications. Their results show that stochastic programming generally aims at decision-making on strategic and tactical levels, while robust and online optimizations focus on operational decision support. Moreover, their results show that scheduling applications mainly apply online optimization. In contrast, applications in the supply chain or energy area most often rely on stochastic programming, often on robust optimizations and rarely on online optimizations. Table 3.1 summarizes an excerpt of their results. Thereby, it omits advanced sub-types of the noted methods and application areas irrelevant for this thesis.

Table 3.1 Frequency of applications using Stochastic Programming (SP), Robust Optimization (RO), and Online Optimization (OO)—reduced representation from Bakker et al. (2019)

Area of application	SP	RO	OO	Uncertainties
Supply chain, production, inventory	●	◕	◔	Demands, capacities
Scheduling	◎	◎	●	Tasks, dates
Energy, electricity, power	●	◕	◔	Demands, prices, resources
Environment, water, air, waste	◕	◎	◎	Demands, emissions, inflows
Traffic, transportation	◔	◔	◕	Travel times, orders, locations
Packing, loading	◎	◔	●	Sizes, weights
Staffing, rostering	◔	◎	◎	Requirements, demands
Projects	◎	◎	◎	Durations, returns

● very frequently / ◕ often / ◔ occasionally / ◎ none

Following the functional requirements described in the last chapter and the characteristics imposed by the development of methods for a decision support system, this thesis focuses on applying a predictive-reactive approach using a flexible job-shop scheduling formulation. Compared to purely reactive approaches, a predictive-reactive setup reduces the number of recourse actions by anticipating dynamics during the scheduling. In comparison to purely preventive approaches, this type of dynamic scheduling reduces the computational requirements (stochastic optimization) and the number of recourse actions implied by the worst-case assumption of robust approaches. Due to the need to mitigate weather dynamics during online scheduling, the proposed approach will follow the general literature's tendency to rely on online optimization. Consequently, the proposed approach applies a specialized form of rolling or receding horizon techniques: the Model Predictive Control described in the next subsections.

3.3 Rolling and Receding Horizon Techniques and Model Predictive Control

MPC constitutes an online optimization scheme originating from the area of control theory. Thereby MPC poses a special case of the more general receding horizon control or the rolling horizon planning in general.

3.3.1 Rolling Horizon Planning

Rolling horizon planning constitutes a technique to cope with time-dependent planning problems that involve uncertainty, e.g., relating to demands. Over the last decade, several authors present literature reviews or applications in several domains, e.g., for the coordination of supply chains (Sahin et al., 2013), transportation planning (Wang and Kopfer, 2015), or production planning (Ponsignon and Mönch, 2014; Ziarnetzky et al., 2018) just to list a few. Rolling horizon planning follows the general idea of separating time into a number of equally spaced time intervals t_i (sampling steps) of length T (sampling step width), denoted as planning periods p_i. At the beginning of each period, the algorithm generates a plan, usually covering several instances, whereas P denotes the number of planning periods included. Finally, it applies the first period before repeating this process for the next sampling step. Depending on the application and associated uncertainty, the sampling step width ranges from hours to weeks, sometimes even months. Some applications use the notion of fixed periods. Thus, the algorithm calculates a plan for several planning periods, applies the first, and fixes a predefined number of future periods for the next iteration. This procedure applies to cases where the planned system cannot react quickly to changes, e.g., in a production context where materials have already been ordered for the second period.

Figure 3.1 shows a schematic example of rolling horizon planning. In each sampling step (vertical axis), the algorithm devises a new plan covering a total of six planning periods (horizontal axis). In each sampling step, the algorithm applies the first part of the plan, i.e., the first planning period, and progresses the time until it reaches the second sampling instance. As this exemplary system cannot react quickly to new plans, the algorithm considers the second planning period of each plan fixed, i.e., this period remains unchanged during the next iteration.

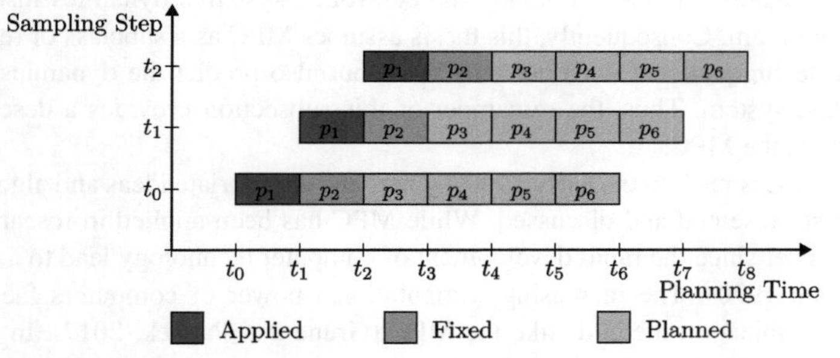

Fig. 3.1 Schematic depiction of rolling horizon planning with $P = 6$

Generally, rolling horizon techniques show advantages in dynamic systems, where disruptions or certain critical values cannot be determined during the planning (c.f. literature review in the next chapter). Depending on the application domain, such values can be the demand, a production rate, or resource availability, e.g., wind, sunlight, or similar. Thus, the algorithm collects current estimations or forecasts at each sampling instance and devises a plan, expecting that these predictions hold. By revising these plans and the underlying predictions at each sampling instance, rolling horizon techniques quickly adapt to dynamics and disruptions within the planned system.

3.3.2 Receding Horizon Control and Model Predictive Control

While literature keeps the general notion of rolling horizon planning comparably general, control theory proposes a very similar control scheme. Thereby, receding horizon control follows the same basic setup and belongs to the class of online optimizations. However, the major differences between rolling horizon planning and receding horizon control are given by the average sampling step width and the degree of formalization: Literature on rolling horizon planning does not provide a lot of formalization and generally aims at re-devising short- to mid-term plans as time progresses. In contrast, literature describing receding horizon control follows a stricter formal framework but, instead of plans, aims to devise momentary control decisions, thus generally applying very short sampling step widths.

According to Grüne and Pannek (2017), the terms *receding horizon control* and *model predictive control* can be used synonymously. Nevertheless, the literature reviews in Chap. 4 show that several authors apply receding horizon techniques without predictive models, tracking the controlled system's dynamics instead of predicting them. Consequently, this thesis assumes MPC as a subclass of receding horizon techniques that does not only track but also predict the dynamics in the controlled system. Thus, the remainder of this subsection provides a description that details the MPC.

MPC traces back to the early 1960s, where the appropriate ideas and algorithms were first presented and discussed. While MPC has been applied in research and practice ever since the rapid development of computer technology lead to a revival in the late 1990s. The increasing computational power of computers facilitated online optimization methods like the MPC (Grüne and Pannek, 2017). In the literature, most MPC applications aim for short-term control of continuous systems to cope with unanticipated or indescribable dynamics. The literature presents several examples of such applications, e.g., production control (Dias and Ierapetritou, 2016; Ivanov et al., 2012), robot motion control or collision avoidance (Ji et al., 2017), energy and building management (Mirakhorli and Dong, 2016), or supplier management (Dastyar et al., 2020a,b).

MPC follows a similar setup, as described earlier. At each sampling instance t_i, with $i \in \mathbb{N}^+$, the controller measures the controlled system's state, usually denoted as state vector $x(t_i)$. The controller then obtains an optimal control sequence $u^\star(t_i)$ for a predefined prediction horizon ($N = P \cdot T$) using the method's name giving predictive system models. Finally, it extracts the first control value $u(t_i)$ of this sequence and relays it as feedback to the controlled system. Figure 3.2 provides an overview of the general control scheme for $P = 3$.

The prediction horizon consists of P many sampling steps, each with a length of the sampling step width T. Therefore, MPC controllers usually solve a finite-horizon optimal control problem, which minimizes the difference between a reference signal $x_*(t_i)$ and the respective, predicted state of the controlled system $x_u(t_i)$. Generally, MPC uses the cost function given in equation 3.1 (Grüne and Pannek, 2017):

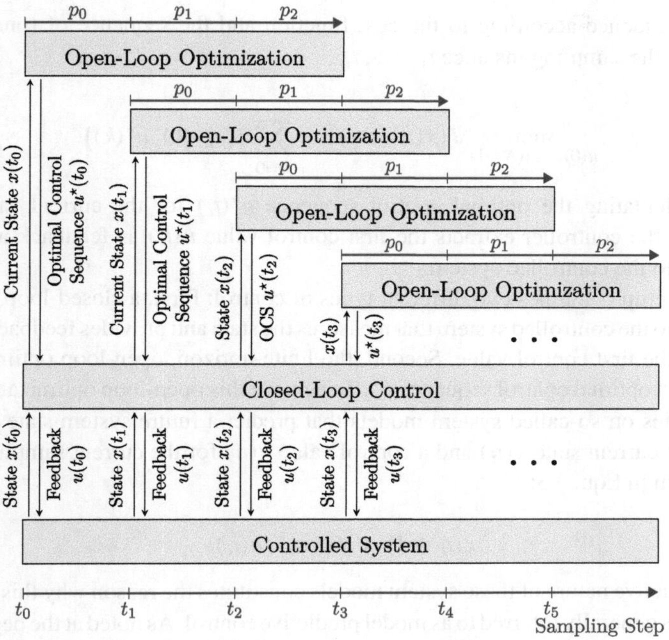

Fig. 3.2 Schematic depiction of the MPC scheme with $P = 3$—derived from Rippel et al. (2019d) ©2019 IEEE

$$\ell\big(x_u(t_i), u(t_i)\big) = ||x_u(t_i) - x_*(t_i)||^2 + \lambda||u(t_i) - u_*(t_i)||^2 \qquad (3.1)$$

Like the reference signal x_*, u_* denotes a reference control, which allows penalizing both the distance from the reference signal and deviations from the desired control value. Therefore, the parameter $\lambda \geq 0$ defines the strength of the penalty on the control. $||\cdot||$ denotes the usual Euclidean norm. This cost function describes a tracking problem, as the optimizer will try to adjust the system state to follow the reference state as close as possible. In cases where the reference signals x_* and u_* are 0, the problem reduces to a so-called stabilization problem (Grüne and Pannek, 2017).

Using this cost function, the MPC solves the finite-horizon optimal control problem given in equation 3.2. In this function, $u^\star$ denotes the minimizing (optimal) control sequence according to the cost function and the sequence of (predicted) states for the sampling instance $t_1, \ldots, t_N$.

$$\min_{u(0),\ldots,u(N-1)} J\big(x(t_i), u(\cdot)\big) = \sum_{k=0}^{N-1} \ell\big(x_{u^\star}(k), u^\star(k)\big) \tag{3.2}$$

After calculating the optimal control sequence $u^\star(t_i)$ for the current sampling instance, the controller extracts the first control value $u(t_i)$ as feedback and forwards it to the controlled system.

This setup combines two different types of control: First, a closed-loop control tethered to the controlled system that measures the state and provides feedback in the form of the first control value. Second, the finite-horizon, open-loop optimization obtains an optimal control sequence $u^\star$. To achieve this open-loop optimization, the MPC relies on so-called system models that predict a future system state $x(t_{i+1})$ given the current state $x(t_i)$ and a control value $u(t_i)$ for the current sampling step t_i as given in Equ. 3.3:

$$x(t_i + 1) = f(x(t_i), u(t_i)) \tag{3.3}$$

The predictive nature of these system models constitutes the reason why this control scheme is generally referred to as model predictive control. As noted at the beginning of this section, the literature shows several applications that follow the general receding scheme to counteract dynamics but do not use the system models to predict them. While these approaches often note to apply MPC, this thesis categorizes them as receding horizon control and, in contrast to, e.g., Grüne and Pannek (2017), differentiates the two terms concerning the predictive characteristics of the system models.

The description above shows that MPC actually constitutes more of a control scheme as a fixed method. It allows the application of a variety of different methods and approaches. For example, Raković and Levine (2019) list several types of MPC, e.g., robust, distributed, stochastic, or nonlinear variants. While the actual implementation and selection of methods for each type differ, they all follow the same general control scheme presented above. This fact highlights the versatility of the MPC scheme.

While the classic setup of MPC describes a tracking or stabilizing problem, literature introduces the terminology of Economic MPC for generalized minimization problems (Raković and Levine, 2019) and first described by Rawlings et al. (2012). In contrast to using a cost function to minimize the distance between the state and the reference signal (c.f. Equ. (3.1)), Economic MPC formulations minimize the cost function directly. These functions commonly refer to economic costs, like material costs, production costs, or energy costs. In the following, this thesis will use the term Model Predictive Control and the abbreviation MPC for both cases, the classic and the economic MPC. Nevertheless, the proposed methodology uses an economic formulation, as it directly tries to minimize times and associated costs.

3.4 Model Transformation

The proposed DSS includes model transformations to decouple the optimization models and the user interface. From a software development point of view, this allows separating the definition and management of scenarios from the model execution code. Moreover, it provides the opportunity to develop more suitable interfaces using different levels of detail for various user groups. For example, these interfaces could focus on the scenario for project managers or the underlying models and methods for simulation or optimization experts. Thus, methods-experts can define and save viable settings for the optimizer, while project managers can more easily define, save, and evaluate various scenarios as needed.

Model transformations fall into the area of the Model Driven Architecture as defined by the Object Management Group in several standards in 2001 and revised in 2004. Therefore, the Object Management Group (OMG) provides several standards, e.g., for transformation languages, baseline model elements, and the overall architecture (Object Management Group, 2020). Literature has established the term Model Driven Engineering to incorporate transformation methods not included in the OMG's standards. According to Kahani et al. (2019), this field of research aims to establish the use of models during the whole process of a system's development, e.g., to simplify communication, design, or compatibility between different stages. This thesis uses methods from the subarea of Model Driven Development (MDD) to facilitate the decoupling of the executable models and the user interface. This subarea focuses on developing transformations from more abstract, descriptive models into models that can be executed directly, e.g., as simulations or optimizations (Kahani et al., 2019). Therefore, the literature differentiates between three different types of transformations: Model-To-Model (M2M), Model-To-Text (M2T), and Text-To-Model (T2M). According to a literature review on applied transformation tools presented by Kahani et al. (2019), various tools and approaches exist for the first two categories, but only a few formal tools for T2M transformations. According to this review, most applications (86%) apply the OMG's EMOF standard to define their models. This standard provides a basic set of model elements, e.g., classes or associations that users can apply to create basically any modeling language. This process of defining a specialized modeling language is generally called meta-modeling, and the language used to define the language is called a meta-model. Consequently, this transformation process generally covers several layers of modeling languages and models.

Figure 3.3 presents an example of such a hierarchy, as later used in this thesis. A general meta-model, e.g., the EMOF, constitutes the highest abstraction level in the left branch. By implementing this so-called meta-meta-model, modelers can define their own domain-specific modeling language. The resulting meta-model defines language constructs available to users to define their respective scenarios (user models). In the context of this thesis, such language constructs could describe vessels, wind farms, or complete installation projects. Therefore, each construct defines which other elements it relates to and which attributes it contains. Consequently, a user can simply add a "Vessel" to their model and define the required

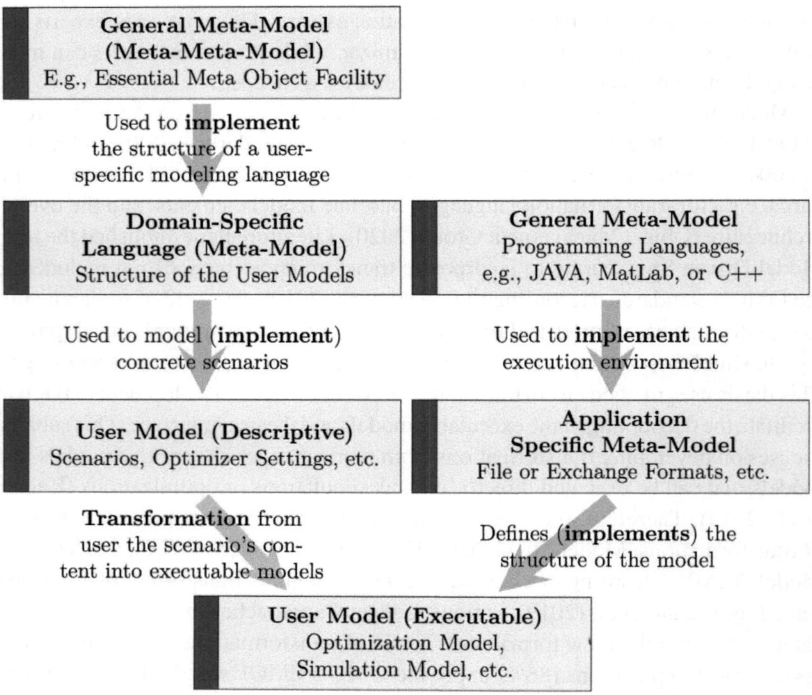

Fig. 3.3 Example of a hierarchy of (meta-)models for Model Driven Development

attributes, like speeds, capacities, or prices. In doing so, the user creates specific, descriptive scenarios. At this stage, these user models are platform-independent, i.e., they cannot be executed directly. Thus, the final step constitutes the actual model transformation into a platform-specific, executable model either by M2M or M2T transformations. The right branch of the figure describes the hierarchy that defines the structure of this executable model. Generally, some execution environment, e.g., an optimizer or simulation tool, will execute the model. This environment defines the executable model's structure, e.g., by providing interfaces or file formats, and, thus, acts as a meta-model for the executable model. Following the hierarchy further up, a general meta-model is given by the respective programming language used to implement the execution environment.

In conclusion, the decoupling of the proposed optimization models (right branch) and the user interface (left branch) requires the definition of a domain-specific modeling language, e.g., using the EMOF, and the implementation of the respective transformation into a format usable by the developed DSS. This thesis will present further details and the selection of actual tools later on in Chap. 6.

State of the Art

4

Following the introduction of various methods and techniques in the last chapter, this section presents the results of several literature reviews on the methodological background and state of the art in the application domain. These reviews aim to identify other approaches in the literature that either apply a similar methodological mix or focus on the same area of application. The results show that no applications could be identified in the literature that satisfy the functional requirements defined in Sect. 2.4. Thus, this chapter closes with a summary of the research gap and the scientific contribution of this thesis.

4.1 Literature Review on Applied Methods

Considering the combination of applied methods, subsection 4.1.1 presents contemporary literature on applications of the MPC scheme in production and logistics. Subsection 4.1.2 provides an overview of applications that explicitly combine receding horizon techniques as a generalization of MPC and MILP techniques. Finally, subsection 4.1.3 provides a review of literature that use MPC as part of a DSS.

4.1.1 MPC in Production and Logistics

This section focuses on literature, which applies MPC for the scheduling in production and logistics. This review covers all articles found via SCOPUS and ScienceDirect, which contains the given combination of keywords in their title, abstract, or keywords, and have been published within the last ten years (2010 and newer).

© The Author(s), under exclusive license to Springer Fachmedien Wiesbaden GmbH, part of Springer Nature 2026

D. Rippel, *Online Decision Support for Offshore Wind Farm Installations*,
https://doi.org/10.1007/978-3-658-49912-9_4

Moreover, this review excluded articles with the keywords energy and chemical in their titles. While the chemical industry applies MPC extensively for controlling operations, its continuous processes differ strongly from those regarded in this thesis. Later on, Sect. 4.1.2 covers articles focusing on the control of energy systems. The search used the following search string:

```
(MPC ∨ "model predictive control") ∧ (production ∨ logistics
∨ installation ∨ erection) ∧ scheduling ∧ ¬ (chemical ∨ energy)
```

The search returned a total of 42 articles published within the last ten years. A first screening of the abstracts excluded 17 articles. The results listed some articles twice, e.g., once as conference contribution and, later on, as an extended version in a journal. Some of the excluded articles provide the requested keywords, but with another meaning, e.g., one article from biology focused on viral (re-)production. Despite excluding the keyword "energy" and "chemical" from the search, 12 of the remaining 25 articles dealt with these areas. Of the remaining 13 articles, one focused on the control of transport systems, four on the control of continuous representations of production systems, three on the control of single production units, and four on the actual control of machining or production operations. The final result presents a survey of MPC applications from a supply chain point of view.

Several articles found in this search do not generate schedules but try to satisfy an externally generated schedule by controlling the system. Thus, these articles focus more on process or production control than on actual scheduling, as described earlier. This control is common in the chemical sector, where factories need to achieve targeted amounts of products using more or less continuous production processes. Accordingly, these and similar articles represent the production system in terms of a flow model, i.e., represent machines in terms of their in- and output rates and their buffers as inventory levels. The MPC then controls the flow rates to meet stable inventory levels and the targeted production amounts. Among others, Jung et al. (2015), Sprodowski et al. (2018), Zhang et al. (2019b) provide examples of such applications. Comparably, Mrugalska and Stetter (2019) present an approach that uses MPC to control the routing of automated vehicles. The vehicles try to satisfy a predefined production schedule, delivering components to stations at or before a fixed time instance. The MPC decides which vehicle takes over transportation jobs between the work stations, primarily regarding the vehicles remaining battery charge and the deadlines provided by the predefined schedule.

Wenzelburger and Allgöwer (2019) present an MPC application for the scheduling of machining operations. Therefore, they represent the production system as a Petri-Net in the form of tasks, orders, and machines. To solve the scheduling problem, the authors transform the Petri-Net into a state-space representation, enumerating all possible transitions at each possible state. The authors propose to use this state-space to solve the complete scheduling problem at each sampling instance by choosing a sufficiently large prediction horizon. The MPC aims to mitigate the influence of new, previously unplanned, orders. Even though the authors denote their approach as MPC, the described approach strongly resembles a standard rolling horizon planning. At each time instance, the approach generates a new schedule using the current state of the system. The dynamics modeled as part of the MPC only describe the shift of tokens within the Petri-Nets state-space, i.e., denoting which machine takes on which tasks, instead of describing expected dynamics of the production system. Lefebvre (2018) proposes a comparable MPC-inspired algorithm to traverse and optimize the tokens' path through Petri-Nets. Gupta and Maravelias (2017) present a similar approach using a state-space representation of a chemical production system. They generate time-indexed schedules for the production units, which they refresh using a standard receding horizon technique and a slightly adapted time-indexed scheduling formulation.

Lin et al. (2018) present a centralized and decentralized MPC-based formulation for the scheduling of potato starch production. The production system consists of two main elements, the factory, requiring potatoes with a sufficiently high starch content, and farms producing and storing those potatoes. The controller includes a model to predict the starch content for units of potatoes, considering the duration and conditions (barn, field, etc.) of their storage. Using a nonlinear model, the MPC controller first determines how many units of potatoes each farm needs to deliver, based on the unit's current, time-dependent starch content. Afterward, the second controller determines how many units to harvest, move to different storage units, or ship to the factory. The authors apply the MPC to predict the units' remaining starch content over a prediction horizon of ten time steps. This content changes over time, depending on the current storage location. Consequently, the controller tries to determine optimal harvesting and storage sequences to minimize the number of units required to satisfy the factory's demand. The results show that the centralized optimization results in a higher starch yield than a decentralized solution, whereas both approaches result in a higher yield than a fixed process. Another agricultural application presented by McCarthy et al. (2014) applies MPC to control the irrigation of cotton fields. The MPC uses weather forecasts and a nonlinear model of the water content to predict daily irrigation needs. In contrast to the first approach, this application does not provide sequences of operations for each day. Instead, it generates a decision on if and how much irrigation each field requires for the current sampling step.

Ivanov et al. (2018) present a review of control theory applications in supply chain management. The authors highlight recent literature focuses on the following areas for the application of control theory in general:

Dynamic feedback-based production-inventory control: Generally, these applications aim to control inventory levels of production processes using a state-space representation of the underlying production system. They usually involve five components: demand forecasting, lead-times, inventory, work-in-progress feedback, and target stock values. In this context, several applications of MPC exist in the literature, as already described above.

Optimal production planning: In this context, the authors list several publications that offer optimization models for scheduling. A brief review of these articles shows that the majority of these articles apply variants of standard scheduling models, as described earlier. Some other articles focus on simplifying the overall scheduling problem by applying a receding horizon and scheduling subsets of available machines, i.e., only those currently idle. The authors list no article explicitly applying an MPC scheme.

Supply chain disruptions: The review present several articles dealing with control-theoretic applications to counteract various supply chain disruptions, e.g., the bullwhip effect or their so-called ripple effect. Furthermore, the authors shortly sketch their approach to analyze the robustness and resilience of supply chains. None of the presented applications applies MPC explicitly.

In conclusion, the authors state that control theory, in general, has become popular in the context of supply chain management and production. According to the author's results (c.f. Table 1 in Ivanov et al. (2018)), MPC in a production context has been applied for the scheduling of continuous processes, particularly to mitigate (demand) forecast uncertainties. In the supply chain context, MPC has been proposed to enable an adaptation to external disruptions, mostly by exploiting the recalculation introduced by the receding horizon. The authors did not encounter applications for their search areas "Scheduling in Industry 4.0 and Cloud Manufacturing" or "Production-Marketing Decisions".

The results of the literature review presented in Ivanov et al. (2018) fall in line with the results of the review presented in this section: While MPC has been applied in the context of production scheduling, most literature follows the same trends and limitations, and, thus, allows to draw the following conclusions:

-1- Generally, MPC is applied to continuous representations of production systems, mostly for the control of chemical and energy-related processes.
-2- Most approaches apply MPC for its iterative character to mitigate uncertainties and dynamics instead of integrating these in their predictive models.

4.1.2 Receding Horizon Techniques using MILP for Scheduling

This section presents the results of a literature review, which aims to find applications that combine receding horizon techniques with MILP models, independent from their area of application. Therefore, we performed a literature search using SCOPUS and ScienceDirect, using the following advanced search keywords:

(MPC ∨ "model predictive control" ∨ "receding horizon") ∧ (production ∨ logistics ∨ installation ∨ erection) ∧ scheduling ∧ (MILP ∨ "mixed integer" ∨ "mixed-integer")

While this thesis explicitly aims to apply MPC as a specialized form of receding horizon techniques, excluding the general class of receding horizon approaches from the search results in a total of only 19 articles found on SCOPUS and ScienceDirect. The presented review also includes the general class "receding horizon" to provide a broader view of this combination of methods. As a result, the search engines returned 30 articles that match these search conditions within their title, abstract, or keywords. The first screening of these articles showed that eight articles did not match the criteria. Instead, these articles explicitly stated to not rely on receding horizon techniques or MILPs, were duplicates of other articles, or used acronyms in another way, e.g., MPC for Multi Product Campaign. In cases of duplicates, older versions, e.g., conference articles with similar content, were removed from the results. Apart from a single article, all of the search results attribute either to the category of energy systems/microgrid scheduling (9) or to production scheduling (12). The later cases usually refer to continuous processes in the chemical industry.

Some of these articles propose a hierarchical application of MILPs and receding horizon techniques. Therefore, they use a standard MILPs optimization to obtain a general schedule and apply receding horizon techniques to adjust these plans if necessary. The applied methods differ by author and include, e.g., classic controllers (Chu and You, 2014, Rossi et al., 2017, Tong et al., 2017), faster to solve linear programs (Sokoler et al., 2017), or rescheduling (Lindholm et al., 2013).

Several authors provide receding horizon or model predictive approaches that use MILP as a baseline in different contexts. A large percentage of these articles focus on the management of energy systems or smart microgrids. The remainder of this section focuses on these articles.

Bartolucci et al. (2019) present a model predictive control approach that uses a Mixed-Integer formulation for the energy demand of a furniture production process. This article aims to balance the energy consumption and production of a net-zero-energy facility in sampling steps of 30 minutes. Thereby, they use the MPC scheme to track the power generation of solar panels and adjust the storage levels of connected batteries (loading, depleting), depending on the process's energy demand.

Shehzad et al. (2019) present an application of a receding horizon technique to control the hydrogen power storage for a wind plant. They model the power generation, the hydrogen storage characteristics, and the resulting cost function using linear constraints. The model aims to determine when the hydrogen storage should switch on, off, or go into standby given a-priori known timelines for the wind power generation and the market demand. Thereby, each state change, as well as the modes on and standby, imply costs. The authors use the receding horizon to track the current power demand and production and incorporate a physics-based model of the underlying storage system's decay. In the given article, the authors use a sampling step width of one hour to stabilize the energy production over a simulated horizon of 24 hours.

Mouli et al. (2019) present a time-indexed MILP formulation for the charging of electric vehicles using a solar power array and connected energy storage. Thereby, they apply an MPC scheme to select when and how a charging station charges a vehicle depending on the mean predicted solar power generation. The article proposes a cost function that, among others, relies on the costs of buying extra energy from the market or offering stored excess energy back to the energy grid. The authors use a sampling horizon of 15 minutes to update the predicted energy production and current state. They use the mean predicted solar power generation for 24 hours as a baseline.

Similar formulations that focus on minimizing micro-energy grid operational costs in terms of energy production, storage, and acquisition can, e.g., be found in Gambino et al. (2016), Holjevac et al. (2018). These formulations use the receding horizon technique to compensate for unknown energy production rates and market prices or apply an MPC scheme to revise decisions in a short-term manner if the forecasts were wrong. Therefore, these approaches usually assume new forecasts every 24 hours and use a small sampling step width of 15 minutes to one hour. Gruber and Prodanovic (2014) provide a slightly different approach by applying two nested MPC schemes: one for the long-term planning with a sampling step width of 1 hour and one for the short-term control with a sampling step width of 5 minutes. Thereby, the long-term plan determines parameters, which the short-term

MPC considers fixed. Again, the authors assume an update to the demand, price, and energy production predictions every 24 hours.

Besides these applications in the energy sector, several authors focus on production-related or other applications: Knudsen et al. (2014) propose a technique to simplify a scheduling model for the activation and routing of shale-gas wells using Lagrangian relaxation. The authors propose executing this scheduling using a receding horizon to compensate for changing demands and changes in the gas field structure.

Baldea et al. (2015) present an implementation of a scheduling-oriented MPC. This approach uses a Mixed-Integer Dynamic formulation to reduce the time required to compute standard nonlinear scheduling models for continuous (chemical) production processes. The results show that the iterative application of this simplified model achieves comparable results to a full scheduling formulation.

Comparably, Vargas-Villamil and Rivera (2000) propose the use of MPC to optimize a semiconductor factory. Therefore, they represent the factory in terms of flow rates and stock levels (continuous model) and propose a two-staged control cycle. While they apply the MPC to determine target flow rates and inventory levels for one shift based on current demands and jobs, a secondary non-MPC controller selects the scheduling policy (dispatching rule) in 5-minute cycles to achieve the set targets.

Another article included in the search results describes an application of the MPC scheme for the routing of pallets in a de-manufacturing plant (Cataldo and Scattolini, 2014). As with the last article, the authors reformulate a Mixed-Integer Linear Dynamic model into a MILP to simplify calculations. Afterward, they apply MPC to obtain routing decisions. Within the literature, MPC applications are common for routing and pathing decisions, e.g., to compensate for unwanted drift or other unforeseen events.

Assimizele et al. (2018) present an optimization for tugboats positioning to achieve maximum coverage in cases of grounding accidents, including the effects of oil-spills. They formulate the cost function to maximize the probability that one of the tugboats can rescue another vessel if an emergency occurs. First, they present a nonlinear Binary Integer Formulation to cover all constraints, e.g., allowed positioning of the tugboats, drifts, and effects of grounding accidents. Second, they linearize this model, obtaining two MILPs, one for the optimizer's upper and one for the optimizer's lower boundary. They apply these models using a receding horizon to adjust the current estimates for the drift approximately every one or two hours, depending on the scenario. Furthermore, the authors state that the inclusion of weather forecasts might improve the results even further. The given article evaluates the proposed approach in several use-cases, stating a possible reduction of socioeconomic costs, e.g., as a result of oil-spills, by over 45%.

In conclusion, this literature review shows the same results as the previous one. Most authors use the MPC to track varying dynamic aspects of the system instead of predicting these. Consequently, this review allows drawing additional conclusions about the methodological state of the art:

-3- System models usually model time-dependent factors of the technical system, e.g., deterioration or decaying energy capacities, and neglect external dynamic influences. Usually, they only track these external factors.
-4- Most literature using Mixed-Integer models relies on nonlinear or dynamic modeling principles. Often MILPs are used to approximate complex models and to reduce the optimization's complexity.
-5- Resulting from the tracking of dynamics, the limited use of MILP, and the focus on control instead of planning, none of the identified approaches apply classic scheduling, as described in Sect. 3.2.

4.1.3 MPC in the Context of DSS

The previous reviews focused on MPC applications in production and logistics or the methodological combination of MPC and MILP as a scheduling method. This section finally presents a literature review of applications that apply MPC as part of or to provide a DSS in production and logistics. Therefore, the review uses the following keywords in SCOPUS and ScienceDirect, again searching the articles titles, abstracts, and keywords:

```
(MPC ∨ "model predictive control") ∧ ("decision support
system" ∨ DSS) ∧ (production ∨ logistics ∨ installation ∨
erection)
```

The search resulted in a total of five articles. None of these articles proposed an actual DSS using an MPC as their baseline. Of these results, the initial screening removed two of the articles. One article used the acronym MPC for "manufacturing planning and control" and did not fall into the scope of this thesis as it does not use model predictive control. The other article does not fall into the categories of production and logistics but healthcare and proposes a controller using MPC to substitute DSSs for respiratory support. A closer evaluation of the remaining three articles full text showed that two further articles do not propose a DSS using MPC:

One of the remaining three articles presents a literature review on methods for precision irrigation in the context of agricultural productions. This article lists distinct DSSs and also includes the MPC-based optimization presented by McCarthy et al. (2014), described earlier in this section.

Gallestey et al. (2002) present an optimization model that optimizes the trade-off between a power plant's remaining lifetime, stressed by turning it on or off, and its profitability, e.g., regarding fuel prices or market demands. As both factors change over time, the authors propose to solve the model using an MPC scheme by dividing the optimization into several equally spaced time intervals and solving the models consecutively. Unfortunately, the authors do not describe how the actual model should work as DSS or if the MPC only aims to solve the optimization problem (open-loop) or also to update the system state (closed-loop) regularly.

According to the abstract, Zheng et al. (2017) propose a two leveled DSS using a steady-state optimization on the first level and a model predictive optimization on the second level. Unfortunately, the remainder of the article is written in Chinese. A translation of the article, and further information given in Cai et al. (2018), indicate that the authors propose to use the first layer to determine a general steady-state for the system using a receding horizon to compensate for external dynamics. The second control level uses a predictive component to propose the optimal operation, based on the current state of the chemical production system or unit, the (new) steady-state target, and the predicted outcome of available operations.

In conclusion, the references found in this search support the already drawn conclusions. MPC is usually applied on a control level to ensure that the system satisfies predetermined targets and is usually applied for its receding horizon to mitigate external dynamics. Moreover, this search results in one additional conclusion:

-6- Basically, no literature exists, which applies MPC to develop DSSs.

4.1.4 Summary and Discussion

The presented approaches obtain a single control for the current instance, i.e., one value for each controlled variable, which is valid until they reach the next sampling step. Afterward, a new optimal control sequence is generated (**-2-**). Particularly in energy applications, authors mostly rely on the receding horizon aspect to adapt uncertainties or dynamics. If the authors apply MPC, they usually only model technological aspects of the energy systems as part of the MPC's system dynamic and only marginally include other dynamic aspects, like weather forecasts or external disruptions (**-3-**).

Production-related approaches described in the reviews demonstrate that MPC provides particular advantages if dealing with time-dependent dynamics in the system. The MPC's predictive system models allow estimating these developments and enable the optimizer to select optimal control sequences. Nevertheless, as before, none of the described approaches uses the generated open-loop plans after their creation (**-2-** and **-5-**). Moreover, most applications found in the literature deal with continuous or fully enumerated state-space representations of production systems. Thus, no application combines classical operations scheduling with an MPC scheme (**-1-**, **-4-**, and **-5-**).

In general, the reviews show a different interpretation of the term scheduling. While the concepts described in Sect. 3.2 aim to generate plans for a specified time horizon or a given number of jobs, applications of the MPC usually aim at an ad-hoc control of the system to achieve specified targets. While some applications use various forms of scheduling during the open-loop, all approaches discard this plan and only apply the first control for the first, fixed time instance. This behavior can be explained as most approaches neglect to model all dynamics and ignore the influence of (forecast) uncertainty. As a consequence, the literature review on the application of MPC in the context of DSS very few results (**-6-**).

4.2 Literature Review on Offshore Applications

While the reviews on the methodological background show that several applications using the proposed mix of methods exist, none of the identified approaches use these methods in the context of DSSs or for planning tasks. This section focuses the literature review from a general review on applications of the methods to a review on approaches within the application domain. Thus, Sect. 4.2.1 presents a literature review on the topic of DSS for offshore logistics, while Sect. 4.2.2 presents applications for the planning, scheduling, and simulation of OWF installations.

4.2.1 Decision Support Systems for Offshore Logistics

This section presents a review of the literature for decision support systems for offshore logistics. As before, the review used a keyword search in SCOPUS and ScienceDirect to identify potential literature. Following, we applied a screening of titles and abstracts to focus on relevant articles, which we then analyzed in more detail. Searching with the following keywords resulted in a total of 24 matches. This search explicitly focuses on offshore applications and does not restrict the results to wind farms, to include applications in the oil and gas area.

```
("decision support" ∨ "DSS") ∧ logistics ∧ offshore
```

The initial screening of these articles excluded 16 articles from the results. Several articles (11) aim at decision support for operations and maintenance for offshore facilities. Thereby, these articles focus, e.g., on the determination of weekly supply and maintenance routes for oil rigs (Borthen et al., 2019) or the maintenance fleet optimization (Sperstad et al., 2017) and simulation (Endrerud et al., 2014) for OWFs. Consequently, these articles mostly focus on fleet optimization and routing. Two articles present technical systems, e.g., for enhanced weather sensors or autonomous maintenance vessels, and three of the articles do not fit the search topic at all. The remainder of this section summarizes the approaches presented in the remaining eight articles:

Leontaris et al. (2019) present an approach to evaluate expert opinions for later use in decision support systems. While this article does not directly relate to the topic at hand, it highlights one of the major problems in the installation of OWFs: While the number of installation projects increases world-wide, this area still lacks a sufficiently large data basis for conclusive evaluations. Generally, only a few experts exist, and most of the project's details remain in the hands of the respective companies.

Barlow et al. (2018) present a mixed-method approach for the planning of offshore installations. Therefore, the authors propose to combine a discrete event simulation with a robust optimization model to enable scenario-based evaluations of installation projects. The approach aims to estimate costs and durations for different project configurations, e.g., starting dates or fleet configurations. The simulation relies on a synthetic weather time-series model. The authors generate the schedule applied during the simulation using a two-stage algorithm. The first stage assigns all open tasks to available resources, trying to balance their expected base processing times. The second stage employs a robust Mixed-Integer scheduling model that provides an optimal schedule assuming the worst possible weather conditions for the respective project duration. Finally, the simulation executes the planned operation sequences and obtains incurring costs and the overall installation duration using generated higher-resolution weather data. The authors state that their approach aims to provide decision support before the project starts, i.e., during the determination of the starting date and fleet mix. Nevertheless, they note that the approach can be applied using a rolling horizon, e.g., when available vessel change.

Devoy McAuliffe et al. (2018) present a collection of optimization and simulation models that resulted from the European LEANWIND research project. In conjunction, these models provide the means to optimize various activities around OWFs from their installation (supply chain and port layouts, production, and transport planning) over their operations and maintenance (fleet management, routing) to their decommissioning. The included optimization models aim to provide static, optimal decisions using a-priori estimates for weather and other stochastic variables. The simulation models use Monte-Carlo simulations to evaluate the incurring costs of specified scenarios under stochastic influences. According to the article, the scheduling model for offshore installations provides indicative plans with a daily resolution, i.e., it estimates how many components can be installed per day using historical weather data.

Muhabie et al. (2018) present a discrete event simulation for the long-term planning of OWFs installations. This approach uses historical weather information to estimate if a vessel can perform operations. Therefore, the authors propose two methods: The discrete approach uses records of a single year for sampling and evaluates each operation against the provided data. The probabilistic approach uses a binning approach to construct a probability for suitable weather windows, which is then used during the simulation to estimate if an operation can commence or not. The approach does not aim to provide actual schedules but to estimate the installation project's duration depending on the selected start date.

Backe and Haugland (2017) present a model for optimizing the fleet size and mix for an OWF installation project. The approach uses historical weather records to identify time windows where operations can be conducted with a daily resolution. Afterward, it generates an estimated schedule for the provided mix of vessels to evaluate incurring project duration and costs. As with the last approaches, this model aims to provide decision support before the project starts. While this approach provides an estimated schedule for all vessels, the focus lies on evaluating the fleet mix.

Leontaris et al. (2017) present a stochastic simulation for the planning of cable installations. Therefore, the algorithm first calculates the monthly probability for the availability of weather windows and applies expert estimations for other disturbances, e.g., missing cables. Afterward, the simulation estimates the installation duration using these a-priori data.

Vis and Ursavas (2016) propose a simulation model to evaluate the effect of different preassembly strategies on the overall performance during the installation. Before starting an operation, the model checks for a sufficiently high probability for a suitable weather window. Therefore, the authors classify weather into three categories and analyze historical weather data to provide a frequency distribution of window lengths. The overall approach executes a process simulation, mainly in terms of storage and transportation capacities, the number of lifting operations, and the overall process duration. Due to the preprocessing of weather data, this approach mainly supports project planning in terms of installation strategies.

Lange et al. (2012) present a network-based process simulation of the supply chain for OWFs installations using Plant Simulation. The simulation aims to derive a long-term master-plan for the installation. It considers resource and component breakdowns, storage shortages, and weather disruptions. Thereby, it relies on randomly generated numbers for these events. The results highlight that the weather uncertainties highly impact the logistics costs during the offshore installation stage. Generally, the proposed simulation aims to provide long-term estimates on a supply-chain level before the actual project starts.

In conclusion, this review shows that only a few works focus on decision support systems in the offshore area. All of the approaches included in the screening highlight the importance of weather dynamics, both for the operations and maintenance area and the installation of offshore facilities. The presented review allows drawing several more conclusions regarding the state of the art in decision support for offshore operations:

-7- Only a few approaches in the literature aim to provide decision support for offshore operations. Even fewer approaches support the installation of offshore facilities.

-8- All identified DSSs aim at providing long-term decision support. The methods rely on (aggregated) historical weather data to provide project managers with a tool to evaluate future projects regarding different aspects, e.g., start dates or fleet mixes.

-9- None of the identified approaches supports live weather data, and, thus, no approach provides decision support for already started projects.

4.2.2 Scheduling of Offshore Wind Farm Installations

While the previous subsection describes approaches for decision support for offshore operations and projects, this subsection provides an overview of methods for the operations scheduling in the context of OWF installations. This review aims to evaluate if current scheduling approaches can be used for the planned DSS regarding the requirements and design features defined in the last chapter. In addition to the articles identified in previous searches, this search used the following keywords to identify a total of 45 articles via SCOPUS and ScienceDirect, published since 2010.

After the initial filtering, only a few additional articles remain. Therefore, this search extends the results by following cited, citing, and related literature.

`scheduling` $\wedge$ `offshore` $\wedge$ `(installation` $\vee$ `erection)`

As there only exists a few literature on the installation of OWFs, e.g., compared to the operations and maintenance area (c.f. previous reviews or Vis and Ursavas (2016)), this extended review resulted in several articles already described earlier. The remainder of this subsection focuses on describing those approaches not yet presented in this chapter, mainly omitting a re-description of simulation-based approaches, e.g., presented in Lange et al. (2012), Muhabie et al. (2018), Vis and Ursavas (2016). Nevertheless, this section shortly repeats mathematical models, focusing on the model's scheduling aspect or approach. Consequently, this state of the art classifies methods presented in the literature into simulation-based and mathematical approaches.

4.2.2.1 Simulative and Descriptive Approaches

The literature includes several simulation-based or descriptive approaches for evaluating offshore installation projects, e.g., the models presented in Lange et al. (2012), Muhabie et al. (2018), Vis and Ursavas (2016). While these models do not aim to provide schedules, the simulations perform sequences of operations. After a simulation run finishes, project managers can use the results to extract possible schedules.

Ait Alla et al. (2017) describe a multi-agent-based simulation model implemented in AnyLogic. This model aims to compare a port-feeder installation concept with the conventional installation concept. Later, Oelker et al. (2018) describe an extension of this model for offshore-feeder concepts. Comparable to, e.g., Muhabie et al. (2018), Vis and Ursavas (2016), the model uses a statistical analysis of available weather windows to decide if a vessel's agent can conduct an operation or not.

Beinke et al. (2017) and Beinke (2018) present a multi-agent, discrete event simulation to study possible cost savings by sharing resources between different concurrent installation projects. The experiments apply real weather recordings to simulate the simultaneous construction of three OWFs using a shared pool of heavy-lift and jack-up vessels. The results show that resource sharing might reduce the required charter time for vessels drastically.

Quandt et al. (2017) describe a simulation model to study the impact of information sharing between different stakeholders in the installation process. This simulation applies sequences of weather classes derived from historical recordings to simulate an installation process. Thereby, the authors compare different scenarios in which stakeholders share the current port capacity, utilization, and weather forecast data or not. The results show that the availability of information, especially current weather forecasts, highly influences the installation project's duration.

Recently, the authors presented another study that combines resource and information sharing into a single multi-agent simulation implemented using AnyLogic (Beinke et al., 2020b). Again, the simulation uses weather classes derived from historical records. The results highlight that resource and information sharing results in decreasing installation times and costs.

Peng et al. (2020a) present a simulation model using Petri-Nets to evaluate the required inventory level and port capacity for offshore installation projects. Later they extend this model by applying simulated annealing to obtain optimized installation cycles (Peng et al., 2020b). The proposed model applies historical data directly from a database, thus, constituting an offline approach.

Apart from simulation studies on the actual installation process, several recent articles provide studies on the effects of the trends noted in Chap. 2, i.e., an increase in simultaneous projects and the components' dimensions:

Cheng et al. (2019) present a Monte-Carlo simulation, based on a Fuzzy Bayesian Network, to estimate the duration of installation projects, depending on fuzzified risk assumptions. These risks, e.g., include weather, material shortages, and engine failures. Therefore, the authors use probabilities for each risk to estimate its effect on the installation process.

Beinke et al. (2020a) recently published a simulation study using AnyLogic that evaluates the future need for jack-up vessels. Therefore, the authors simulate the installation and decommissioning of already planned OWFs in Germany's Northern and Baltic Sea. They apply the simulation to a series of historical weather records between the years 1980 and 2010 to obtain realistic estimations for the duration of these projects. The results show that vessel availability can easily become a bottleneck in the future. This result highlights the need for advanced tools to optimize the use of jack-up vessels.

Oelker et al. (2020) present another multi-agent-based simulation study that focuses on evaluating capacity requirements at the base port. While this study does not explicitly target the installation process, the model includes port-side processes, resources, and capacities. The study shows that port-side resources may become a bottleneck in the future due to the increasing number of simultaneous installation projects, newly introduced decommissioning projects, and the increasing dimensions of components.

Besides simulation models, literature also offers several mathematical models that aim at providing duration or cost estimations but do not incorporate scheduling. Generally, these models rely on discrete weather assumptions or neglect these dynamics in general. As these models do not fall into this thesis's focus, this section does not introduce these descriptive models or cost models in detail. As an example of this class of models, (Sarker and Faiz, 2017) describe a cost model for the transport and installation costs of offshore wind farms. The model does not regard weather dynamics but aims to minimize the expected costs in terms of the number and capacity of employed vessels and selected preassembly concepts. Therefore, the model consists of various cost parameters, e.g., capital costs for vessels, interest rates, or return-of-investments for generated power.

While literature shows several simulation-based approaches, these approaches show disadvantages considering the requirements described in the last section. In summary, the following conclusions can be drawn as a result of this literature review:

-10- None of the identified simulation models includes an optimization to select operations intelligently. They all apply predefined rules and simulate the effects, e.g., of weather dynamics or installation concepts on the process.

-11- While several simulation models use historical weather recordings, no model supports the use of forecasts.

4.2.2.2 Optimization-Based Approaches

Scholz-Reiter et al. (2010) propose a combination of a precedence-based job-shop scheduling model with a multi-periodic production formulation to obtain daily schedules. Due to the high complexity of the formulation, they later propose a heuristic solution approach to solve larger instances of the formulation (Scholz-Reiter et al., 2011). The approach uses randomly generated sequences of classified weather windows. Therefore these articles consider windows to last for one to three days and represent either good, moderate, or bad weather. Each class restricts which operations a vessel can perform during this period. In recent years, Ursavas (2017) extended this model to allow for probabilistic weather classes. Therefore, the authors apply a sampling-based stochastic optimization to generate plans for probable scenarios. Nevertheless, the authors do not modify the baseline formulation. This approach mainly supports the strategic or tactical levels and seems unsuitable for online scheduling due to its low daily resolution. Moreover, these articles apply randomly generated sequences of weather windows with a minimum duration of one day. As described earlier, operations usually take approximately two to three hours, rendering these time frames too wide for operative decision support.

Ait Alla et al. (2013) present a model formulation for aggregate scheduling. The concept behind these aggregate plans shows similarity to the indicative plans described in Devoy McAuliffe et al. (2018): The resulting plan does not list single operations but estimates the number of installable components per time slot.

Therefore, their model uses a slot-based scheduling formulation with a slot size of twelve hours. The approach relies on classes of weather windows (good, bad, etc.) to estimate how many operations of a given type can take place within a slot. In terms of operations, the authors regard the installation of foundations and top-structures. The article presents an application to evaluate different fleet mixes. This kind of formulation mainly aims at tactical to strategic support by estimating how many components a fleet of vessels could install in a given period. The current approach does not support forecasts and relies on discretized weather information derived from historical weather records.

Kerkhove and Vanhoucke (2017) present a Mixed-Integer formulation to determine optimal times to commission or decommission jack-up vessels. The model focuses on costs, e.g., for activities but also for contracting, chartering, and commissioning vessels. It minimizes these expected costs through optimal gate-times for each vessel, derived by calculating a schedule for each vessel. The formulation uses a time-continuous representation and schedules all required activities. The model uses a set of conditional probabilities to represent the evolution of weather influences in discretized classes of four hours each. The authors analyzed historical weather recordings of the Belgian coastal region to obtain these probabilities. The presented approach aims at supporting decision-makers on a strategic or tactical level. The authors state that project managers need to reserve vessels up to two years in advance. Nevertheless, costs differ, depending on a vessel's state. The underlying scheduling algorithm does not support weather forecasts and, as a result of its time-continuous formulation, requires to schedule all available operations. Consequently, the formulation seems unsuitable to provide online support.

Irawan et al. (2017) use the same representation of weather classes as (Scholz-Reiter et al., 2010) to propose a bi-objective scheduling model. The model aims to determine the optimal trade-off between costs and installation times. Comparable to the approach presented by Scholz-Reiter et al. (2010), the authors propose to use a daily resolution for their slot-based formulation but try to determine several trips for each vessel, each consisting of one to several daily slots. The authors recently applied the same model for the decommissioning of OWFs, assuming that the operations and restrictions remain identical but reversing the process chain (Irawan et al., 2019). This approach shows the same drawbacks in terms of possible online scheduling as the approach presented by Scholz-Reiter et al. (2010). The daily resolution and the quite abstract representation of weather windows do not allow for an accurate representation of weather dynamics required for online decision support. Thus, this approach, again, aims at the tactical/strategic level.

As noted in Sect. 4.2.1, Barlow et al. (2018) present a mixed-method approach for the planning of offshore installations. This approach includes a robust Mixed-Integer scheduling model already proposed in Tezcaner Öztürk et al. (2017). The robust formulation uses a-priori information on task-vessel assignments and precedence relationships, e.g., foundation before the tower, to obtain a sequence of tasks for each vessel. Thereby, the formulation uses a robust, worst-case assumption for each task's duration. While the original proposal does not describe how the task durations should be estimated, Barlow et al. (2018) apply a synthetic time-series model. The worst-case assumption of this model renders it unsuitable for an online-scheduling, as the model would tend to overestimate the duration and induce a high number of recourse actions. Moreover, the synthetic time-series proposed in later work again aims at strategic or tactical support by providing expected values instead of using current forecasts or data.

The review described above allow to draw the following additional conclusions:

-12- All of the identified scheduling models rely on preprocessed and a-priori known weather information. None of the identified approaches supports the use of current weather data or forecasts.

-13- In consequence, all of these approaches aim at tactical or strategic levels or require frequent reapplications to handle forecast uncertainties.

4.2.3 Summary and Conclusion

The results of this literature review within the application domain support the statement of Vis and Ursavas (2016) that only a few models for the scheduling or decision support of offshore installations exist (**-7-**). All models rely directly on historical data or aggregations of these. None of the proposed approaches supports the use of current weather measurements or forecasts. This accounts for current literature on DSS (**-9-**), current simulation models (**-11-**), and also for current scheduling-based approaches (**-12-**).

Simulation-based approaches generally aim to evaluate specific strategies against a predefined scenario, e.g., by simulating different fleet mixes for several years from historical records. Unfortunately, all applications identified in the literature use a predefined set of rules for the choice of operations, i.e., the vessels always follow the same sequence of operations, whereby the influence of weather conditions mainly delays these operations. None of the simulation-based approaches includes an optimization component supporting decision-makers in their daily (online) scheduling tasks (**-10-**).

In terms of scheduling-based approaches, the same restrictions in terms of current weather measurements and forecasts hold. All of the identified approaches assume deterministic or stochastic weather conditions already known when creating the schedule. While some authors propose sampling-based, stochastic optimizations, most authors apply their tactical or strategic decision support approaches by providing best estimates using historical data. Consequently, all of the described approaches need to schedule all operations, i.e., the complete installation project, rendering an integration of forecasts impractical.

In conclusion, it can be stated that none of the approaches proposed for the scheduling of offshore installations aims at decision support on an operative level by proposing means to handle forecast uncertainties or current measurements (**-8-**, **-13-**).

4.3 Research Gap and Contribution of this Thesis

This section summarizes this chapter's results and presets the resulting research gap, the justification and selection of methods for the proposed DSS, and finally, the theoretical contribution to the state of the art.

4.3.1 Research Gap from the Area of Application

The literature review in the application domain shows that only a few articles deal with the installation of OWFs or propose methods for decision support in offshore installation logistics. Thereby, most of the articles (eleven out of sixteen) that aim at providing decision support for offshore logistics target routing decisions for the maintenance of already installed facilities. All models, both for the scheduling of offshore operations and decision support in offshore logistics, aim to provide decision support on the strategic and maybe tactical level (offline scheduling) but do not allow the use of current measurements or short-term weather forecasts for operative online scheduling. Consequently, this thesis aims to fill the gap by providing a methodology that specifically targets the online scheduling aspect while still providing capabilities for offline applications. Due to the lack of articles explicitly describing DSS for offshore operations, the literature reviews do not indicate how the user interfaces for the proposed methods should be designed. Thus, it can be expected that current approaches require process experts to have deep insights into the underlying models' workings and structures. Consequently, this thesis tries to target this topic by decoupling the user interface and the actual models using model transformation. Thus, a user interface can be developed separately from the models.

4.3.2 Summary of the Justification and Selected Methods

The proposed methodology aims at a predictive-reactive approach to provide a methodological basis for a DSS for offshore installations. As described above, this class of approaches aims to reduce the number of required recourse actions and computational times by estimating the extend of dynamics within the scheduled system. The proposed scheduling formulation follows a flexible job-shop scheduling model as vessels can perform an overlapping set of operations. The proposed methodology applies a specific form of receding horizon planning, the Model Predictive Control, to achieve reactivity. This control scheme consists of two control cycles: One closed-loop control, relying on measurements of the controlled system, and one open-loop optimization that uses so-called system models to estimate the controlled system's development. The open-loop allows deriving an optimal control sequence using weather forecasts. Therefore, the proposed approach requires methods to estimate weather dynamics based on forecast data as part of the predictive part of the scheduling and the open-loop system models. Moreover, MPC reevaluates (measures) the state of the system and the derived plans in predefined intervals. Consequently, the approach uses a time-indexed scheduling formulation, as it allows explicitly defining the planning horizons. Compared to precedence-based formulations, the optimizer does not need to schedule all operations, even if it needs to exceed the forecasting/prediction horizon. The latter would result in unreliable or even infeasible plans if the uncertainty becomes too large. Finally, the DSS includes a simulation component to mimic weather measurements and forecasts using historical data. This simulation component enables offline scheduling, required for higher-level planning tasks. It allows for obtaining schedules from historical or anticipated weather conditions, e.g., capacity or transportation planning. The proposed DSS satisfies all of the characteristics identified at the beginning of this chapter by combining these methods:

Specificity: The methodology aims at online scheduling of offshore operations, both using historical or current data. Section 2.4 describes the involved processes.

Adaptivity / Repeated use: The use of MPC requires a description of the process's current state and the process's general characteristics. By exposing these parameters to the user, the methodology achieves high adaptability considering settings and "what if" analyses. The incremental scheduling achieves frequent reapplications, depending on the state and current forecasts.

Interactive support / Accessibility: As described above, the methodology aims to provide a high level of adaptability for the user and, thus, for interactive decision making support. In terms of accessibility, the methodology employs model transformations to detach the actual optimization models from the user interface. This detachment allows a high degree of customization for specific user groups.

Monitoring policy: The use of an MPC scheme requires the definition of the system's state and sampling and planning horizons. These values clearly define a monitoring policy as they describe the system to the degree that allows closed-loop control.

Intervention and control policy: The MPC scheme, combined with the MILP-optimization, provides a suitable control policy. Thereby, the DSS will derive a new plan whenever it registers a disturbance.

4.3.3 Methodological Contribution and Research Questions

The literature reviews on the applied methods draw a similar picture as the review on the application area. While authors applied each method for various tasks, the combination and interpretation of methods proposed in this thesis are quite unique. Current literature does not contain any approaches providing online decision support for the scheduling of offshore operations. Thus, this thesis proposes to combine an MPC scheme with MILP optimizations for the scheduling. Thereby, this thesis proposes to use the MPC to generate schedules for several days based on weather forecasts. In contrast, literature only obtains short term, momentary control decisions using MPC. In consequence, literature shows a different understanding of scheduling, as proposed in Chap. 3. This thesis interprets scheduling as the derivation of operation sequences for a set of resources. The literature identified during the reviews commonly interprets the term scheduling in a sense to render control decisions, e.g., on when to turn a controlled resource on or off. Moreover, literature applying MPC generally uses system models to describe system-internal dynamics, e.g., the deterioration or loss of battery capacity, while ignoring external dynamics, e.g., weather or changes in the order composition. Most articles aim to mitigate these external dynamics by relying on MPC's receding horizon. In contrast, this thesis tries to incorporate such external dynamics as part of the predictive system models to reduce the number of required recourse actions. The combination of these two methods results in the following research questions considering the feasibility and performance of this combination:

RQ1 Is it possible to apply Model Predictive Control to the online scheduling of offshore operations in a planning context (not just control)? Which methods and models does it require to act as DSS?

RQ2 Does a combination of receding horizon techniques, particularly Model Predictive Control and Mixed-Integer Linear Programming scheduling models, provide advantages over classical scheduling approaches, e.g., stochastic or robust optimizations?

The problem decomposition in Sect. 2.3 shows that offline scheduling of operations constitutes a requirement for other higher-level planning tasks, e.g., planning transportation and capacities. Thus, the proposed methodology will include a simulation component to generate schedules for expected or averaged weather conditions. This results in the following research question:

RQ3 Is it possible to apply the proposed scheduling algorithm to historical weather records to achieve suitable approximations for long-term planning tasks? Which additional preprocessing does such an application require?

Finally, the literature shows that DSSs require targeted user groups to alter and modify the scenario to perform a scenario-based analysis of different settings. Therefore, these users should not need to be experts in the underlying models and mechanics. Thus, the proposed methodology aims to use model transformations to decouple the user interface from the actual models and methods. While the actual interface design will not be the focus of the thesis, it will try to answer the following research question:

RQ4 To which extend is it possible to use model transformation and code generation to provide project managers with a detailed and intuitive interface for the mathematical scheduling algorithm? Can other planning tasks be integrated into such an interface?

Design of the Decision Support System 5

Following the description in Sect. 2.3, the planning of offshore wind farms comprises several stages, ranging from strategic tasks, like the network and capacity planning, down to operative tasks, like the scheduling of vessels and operations. While this thesis focuses on providing a DSS for decision support on the operative level, it also evaluates if, and under which assumptions, the underlying methodology supports higher-level tasks as well. Planning tasks on different levels rely on each other, either by requiring estimations from lower-level tasks or by providing constraints for theses (c.f. Fig. 2.7). For example, the strategic capacity and transport planning require estimations on the number of installable OWTs. The other way around, these tasks can limit the installation's performance if e.g., their results underestimate the required storage capacity at the base port or if resupply cycles become inefficient.

Consequently, this chapter first describes the MPC-based scheduling methodology that serves as a foundation for the DSS in Sect. 5.3. This description includes the definition of the state and its measurement, methods used for the predictive system models, and the MILP optimization model. Afterward, it introduces customized models for higher-level tasks (Sect. 5.4) and the decoupling of the actual mathematical models from the user interface using code generation and model transformations (Sect. 5.5).

Finally, the general concept includes a simulation component to mimic the real-world system. Unfortunately, this thesis cannot rely on process data from a real-world installation project and, thus, applies a simulation of the processes instead. This simulation uses some of the methods proposed as part of the online scheduling methodology, e.g., to calculate the duration of operations using historical weather data. Instead of using current weather measurements and forecasts, this simulation directly uses historical weather records to provide a realistic setting for the evaluation of the approach. As the scheduling methodology requires a real-world system to evaluate its plans and to obtain new measurements, this section presents the simu-

© The Author(s), under exclusive license to Springer Fachmedien Wiesbaden GmbH, part of Springer Nature 2026

D. Rippel, *Online Decision Support for Offshore Wind Farm Installations*,
https://doi.org/10.1007/978-3-658-49912-9_5

lation component as part of the online scheduling approach. Nevertheless, practical applications would not require this simulation and retrieve information from the real-world instead.

5.1 Model Assumptions and Characteristics

The overall planning problem consists of several interconnected planning tasks. Thus, the models proposed in this chapter cover some aspects that link to other planning tasks while they omit other aspects that are subject to future research. Generally, this thesis focuses on scheduling offshore operations and additionally covers the capacity and transport planning for components to the base port. This thesis omits other topics, like the personnel and network planning, to keep the focus on operations scheduling. Thus, the following assumptions and limitations hold for the presented models:

- The online scheduling only considers the actual installation part of the given supply chain. The thesis proposes separate models for capacity and transport planning as examples for higher-level planning tasks.
- Due to the practical requirement to finish an installation after jacking-up, the presented models use so-called aggregate operations. Thus, the scheduling model aggregates all installation-related operations into a single operation with a highly variable processing time. Thereby, the different weather restrictions of the included base operations and the current (predicted) weather conditions determine this processing time for each instance of the aggregate operation.
- Similarly, the models aggregate components into full sets of components, each consisting of a tower, three blades, and the nacelle and hub. While this assumption simplifies the consideration of storage areas for capacity planning, it constitutes a viable simplification for the scheduling. As a vessel needs to finish constructing an OWT after jack-up, partial loading scenarios are not viable in practice. Similarly, this simplification allows considering the resupply of components to the base port in terms of resupply cycles. Thereby, each cycle consists of a predefined set of round-trips. Each trip fetches some of the components from the production port. Once the HLV completes the cycle, i.e., all round-trips, a given number of complete sets becomes available at the base port. The scheduling model does not consider partial deliveries, e.g., just of towers.
- As in practice, this thesis only considers the so-called season from April to October. Months outside this season generally do not provide suitable weather conditions to perform offshore operations efficiently.

- If not stated otherwise, a single scenario considers all employed HLVs and installation vessels equal. Nevertheless, the implementation allows modifying the capacity of single installation vessels. Moreover, it automatically adds a minimal cost (generally 1% of the base cost) to each additional vessel's operations to reduce the model's symmetry. This additional cost ensures that the optimizer cannot exchange plans for vessels without incurring different costs and provides an additional gradient for the optimization.
- In its current state, the models assume that the number of used vessels remains constant over the scenario. While it would be possible to "block" installation vessels for a specific time, the models do not include mobilization and demobilization costs.
- The scheduling model allows limiting the number of available loading bays or cargo bridges at the base port to allow evaluating the influence of limited port side resources. It considers this number fixed for each scenario.
- While the problem decomposition shows that crew management plays a vital role in the planning of offshore projects, the proposed models do not cover this aspect to keep the complexity manageable. Consequently, the models assume a permanent availability of personnel at any time. Additionally, they do not cover dedicated personnel costs for night shifts or weekends but consider personnel costs as part of the variable operating costs.

5.2 Design Targets and Implementation

In comparison to the state of the art in scheduling models for offshore installations and general MPC applications, the proposed methodology shows four characteristics that differentiate it from state of the art methods and applications. First, it uses a time-indexed formulation to allow for higher flexibility with the integration of the MPC. This type of formulation provides several benefits as it allows modifying prediction and planning horizons instead of scheduling all available operations. Second, in contrast to usual applications of the MPC, this approach follows a softer definition of the sampling step width T. Commonly, this parameter determines when a new MPC-iteration occurs, resulting in a new control value or feedback. Instead of obtaining a single value for the feedback, the proposed methodology aims to devise plans covering several operations. Consequently, this approach uses a more variable notion of sampling instances and, while still trying to achieve a specific sampling step width, applies an "on-demand" mechanism to trigger new iterations. Section 5.2.1 provides a detailed description. The third characteristic involves the aggregation of offshore operations. While models in the literature either use a single operation for the installation with a single set of weather restrictions or include all installation operations separately, the proposed approach uses an aggregate notion. Therefore it aggregates all "basic" operations into a single one for the optimization but uses their individual restrictions to estimate the installation's duration. Section 5.2.2 describes these aggregations in more detail. Finally, the last characteristic estimates the duration of offshore operations, considering forecast uncertainties. Approaches in the literature usually either rely on ignoring this uncertainty, robust assumptions, or on sampling-based solution approaches. The proposed methodology aims to include a predictive approach to estimate the duration of offshore operations to mitigate the disadvantages of these assumptions. Section 5.2.3 provides a description of this characteristic and the reasoning behind it.

5.2.1 Planning and Prediction Horizons

In contrast to typical MPC applications, the proposed methodology does not only obtain a single control value, i.e., it does not choose a single operation for each hour the algorithm executes. Instead, it derives a short-term plan comprising several operations and conforms to the selected planning horizon T. The integrated open-loop optimization obtains longer plans for a prediction horizon of $N = P \cdot T$ to maintain the advantages of mid- to long-term plans. However, it only applies the first planning horizon to mitigate increasing uncertainties. This shift from a pure control problem to a planning formulation aims to provide operatives with a higher degree of plannability and reliability by utilizing the MPC's inherent predictive capabilities. In terms of a DSS, this reduces the number of optimizations. Instead of performing an optimization every time step (hour), the proposed methodology only executes optimizations whenever necessary, i.e., when no operations remain in the current plan or if the plan fails. In this context, the methodology considers a plan failing if the operation's planned durations diverge from the time it took to perform the operation in the real-world system.

As noted, the chosen time-indexed formulation allows users to choose the length of the planning and prediction horizons, effectively instructing the algorithm on how many hours to include in the derived plan. Nevertheless, the resulting plans can differ in their length, softening the notion of the sampling step width if compared to the usual interpretation of the MPC. For example, if process planners choose a planning horizon of one week, the algorithms will result in a plan where no operation starts after one week. On the one hand, operations may exceed this one-week limit. On the other hand, the resulting plan can be shorter, depending on weather conditions and the cumulative operations durations. In consequence, the methodology will trigger an optimization "on-demand" with a targeted sampling step according to the planning horizon.

5.2.2 Aggregate and Basic Operations

As described in the state of the art in Table 2.2, the installation comprises several operations, i.e., the loading of components, sailing, positioning within the construction field, jack-up and -down operations, and the offshore assembly of towers, nacelles, and blades. While each operation presents its distinctive weather limits, vessels need to follow a fixed operation sequence during the actual installation. The proposed scheduling model exploits this fixed sequence to reduce the number of operations that it needs to schedule by using so-called aggregate operations. The proposed methodology first estimates the weather-dependent duration of each basic operation, e.g., installing a tower or nacelle, and then summarizes these durations to form aggregate operations. In consequence, this aggregation reduces the number of considered operations during the scheduling from nine to four: *load components*, *move to port*, *move to site*, and *install OWT*. Thereby, it still regards each basic operations' distinctive weather restrictions. While summarizing fixed order operations to a single one is comparably common in scheduling, current literature does not provide any comparable approaches that regard weather- or, in a more general sense, time-dependent characteristics of the included sub-operations.

5.2.3 Weather Forecasts and Duration Estimation

As the proposed methodology provides decision support on an operative level, it processes current weather measurements and weather forecasts. Unfortunately, this type of forecast always includes an increasing uncertainty the further it predicts into the future. Consequently, the methodology includes methods to cope with these uncertainties.

The state of the art shows several ways to manage these uncertainties. While most MPC-applications neglect this type of uncertainty and only rely on the MPC's iterative nature to mitigate its influences, approaches for the scheduling of offshore operations usually apply robust or stochastic approaches. In terms of providing operative decision support, all of these approaches show distinctive drawbacks. A robust, i.e., worst-case assumption or neglecting these uncertainties might result in a very high number of recourse actions, as the devised plans will quickly diverge from the real-world's weather conditions. In turn, stochastic approaches rely on sampling the weather conditions and provide several plans for the most likely conditions. This approach requires the computation of several plans, which might quickly become unsuitable for operative decision support. Scheduling itself constitutes an NP-Complete problem, taking long times to solve. Moreover, as the proposed methodology includes the hourly wave height and wind speed as stochastic values, the search space for possible combinations grows quickly. For example, fully precomputing all possible plans for one week (168 hours), considering only three samples per hour and variable, results in a total of $3^{(2 \cdot 168)} = 2.055 \cdot 10^{160}$ possible combinations. Consequently, all of these approaches would highly reduce the plans' reliability as they either tend to fail (robust, neglecting uncertainties) or as several plans need to be computed and combined ad-hoc (stochastic).

Therefore, the proposed methodology proposes a new approach to handle forecast uncertainty. First, it assumes forecast data to become increasingly uncertain with an increasing prediction period, as later described in Sect. 5.3.2. It encodes this uncertainty directly into the forecasts to render it available to further processing. Second, the methodology includes approaches to estimate the duration of offshore operations, given the operation's weather restrictions and the modified, current forecast. Section 5.3.3 further details this step.

5.3 Methodology for Online Scheduling

Generally, MPC separates the closed-loop into three distinct stages: The measurement and processing of the real-world system's current state, the finite-horizon optimization, and finally, the control extraction and application. Figure 5.1 shows the major steps of the methodology sorted according to these MPC stages.

During the measurement stage, the methodology's algorithm first updates its internal state by obtaining current values, e.g., for the position and state of vessels and a current weather forecast. The algorithm first updates all required dependent values within its state to react to structural changes within the scenario, e.g., if new vessels become available or have been exchanged or if travel times have been updated externally. Section 5.3.2 provides an overview of the state model and the involved calculations. In the second step, the algorithm estimates the duration of offshore operations using the current forecast and the operations' weather restrictions. As this step involves using dedicated methods, the methodology considers it a separate step described in Sect. 5.3.3. The methodology includes this estimation to handle forecast uncertainties induced by the operations' weather dependencies. The measurement

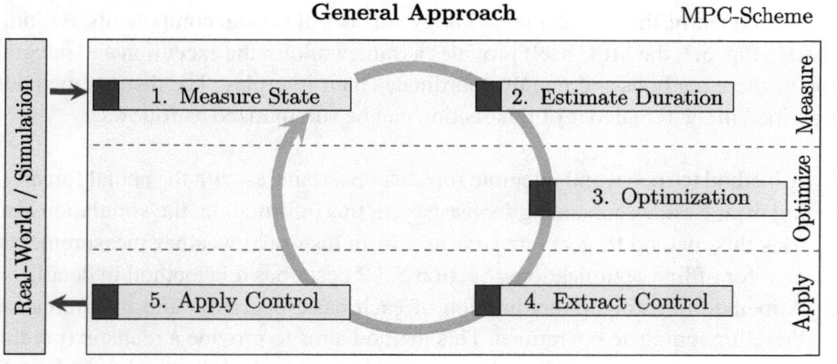

Fig. 5.1 Steps of the methodology ordered according to the MPC scheme—adapted from Rippel et al. (2019b)

stage results in an updated state that the algorithm then uses in the finite-horizon open-loop optimization.

The second stage uses the current state and the optimization model presented in Sect. 5.3.4 to generate an optimal schedule for the given prediction horizon N as a multiple of the selected sampling step width T. The model uses an hourly resolution and the described aggregate operations. The hourly resolution serves as a trade-off between the optimization problem's size and a sufficiently high level of detail to prevent excessive waiting times or aggregation errors. Moreover, in most cases, weather and forecast data provides an hourly resolution as well. As a result, this stage calculates an optimized plan for the desired prediction horizon that the methodology processes in the last stage to obtain the desired plan length.

Accordingly, the third stage consists of two main steps. First, the algorithm extracts plans for each vessel with a duration of not more than T hours. Using these plans, it determines the next planned sampling step as the shortest of these plans and already notes each plans' duration as part of the state. Thereby, it ensures that the next planning cycle will not reschedule vessels that still follow a valid plan, even if one vessel fails its plan and issues a premature rescheduling cycle. The final step conveys the plans for the next planning period to the real-world system or its simulation. While marginal in this thesis, real-world applications could require modifying the obtained plan to specific formats or additional verification until the DSS could release these plans. Therefore, the general approach includes this step for future extensions.

Consequently, the overall methodology consists of several components. As indicated in Fig. 5.1, the MPC itself provides a framework for the execution and integration of these methods and mainly coordinates their interplay. The distinct methods described in the remainder of this section can be summarized as follows:

1. A method to model and integrate forecast uncertainties with the actual forecasting data. Besides enhancing forecasts with this information, the simulation also uses this method to generate forecasts from historical weather measurements, e.g., for offline optimizations. Section 5.3.2 describes this method in detail.
2. A method to estimate the duration of each basic operation and to summarize these for aggregate operations. This method aims to provide a realistic (predictive) estimation of operation durations and tries to avoid the need for robust or stochastic scheduling approaches. Section 5.3.3 provides a detailed description of two possible implementations.
3. The actual scheduling model and solution method to calculate the finite-horizon optimal schedules during the MPC's open-loop. Section 5.3.4 presents the proposed model. This thesis uses a standard implementation of a Branch-and-Bound algorithm to solve the model. This algorithm provides the advantage of incrementally solving the model until it achieves an optimal solution.
4. An approach to extract suitable short-term plans that mostly conform to the selected sampling step width for each vessel. This approach also includes the definition of when the DSS considers a plan failing and provides mechanisms for synchronizing plans across several planning iterations. While this approach does not qualify as a distinct method, it summarizes the monitoring and intervention policies required by any DSS. Section 5.3.5 presents the details of these policies and provides a suitable interpretation.

While the remainder of this chapter presents possible implementations and interpretations for each of these components, practical applications using this methodology could exchange or adapt each of these methods separately. For example, this thesis proposes a function to model and integrate the forecast uncertainty into the provided forecasts. This function could be replaced with information relating directly to the forecasts or using practical experience in practice. Moreover, this thesis presents two distinct methods for the duration estimation, demonstrating that different approaches can be chosen for each noted method.

5.3.1 Simulation of the Real-World System

While this thesis does not consider the simulation part of the overall scheduling methodology, it serves two distinct purposes. First, in the context of this thesis, it substitutes a real-world system by simulating the installation process using historical weather records to enable a realistic evaluation of the methodology. Second, the offline optimization tasks, e.g., the base port capacity planning, require estimates on how many turbines could have been built in a given amount of time. Therefore, this component simulates an installation using different weather conditions, either directly taken from historical records or using aggregates of such records. Consequently, it constitutes a vital component for the DSS's higher-level offline optimization tasks.

First, it verifies plans against measured weather conditions and reports their state of execution to the scheduling methodology. Therefore, it compares the planned duration of operations, including forecast uncertainties, with the actual duration that it calculates using historical weather conditions. The simulation applies the same approaches described in Sect. 5.3.3 3 but without any uncertainties to calculate the operations' real duration. Generally, this process compares the operation's weather restrictions with the provided conditions to determine the waiting time before the operation finishes. The simulation then reports any deviations back to the scheduling methodology, which decides if it considers the plan failing or if it proceeds as planned (c.f. Sect. 5.3.5). Regarding the online scheduling's evaluation, the simulation uses historical weather records. In contrast, capacity planning uses averaged weather data, including several years of historical data, to achieve a higher generalization.

Second, the scheduling methodology requires weather forecasts to calculate its schedules. Consequently, the simulation component provides such forecasts as part of the regular state update. Therefore, it applies the methods described later on in Sect. 5.3.2.2 to generate forecasts from historical weather recordings. Basically, the described approach uses the records as a baseline to generate broadening intervals of probable weather data, which grow depending on the forecasting period. While the simulation uses the recordings directly as ground truth, the generated forecasts only include these intervals to model forecasts' uncertainty. This process ensures that the scheduling methodology does not have access to discrete recordings but only to these uncertain forecasts.

5.3.2 Description of the System State and State Update

The first stage of the methodology starts with the state update. The methodology triggers this update every time it needs to provide a new plan, either because an old plan has finished or failed. The state update requests various information from the real-world system, or, respectively, the simulation component. The state categorizes this information in three major classes: optimization parameters influence how the methodology itself works and, e.g., control the sampling step width or the time available for optimizations. While these parameters usually remain static, their inclusion in the state update allows process planners to react flexibly to changing conditions, e.g., if they need a new plan fast or if they want to change the length of generated plans on-the-fly. The second class comprises so-called scenario parameters. Like the first class, these parameters remain mostly static and describe the actual planning scenario, e.g., the number of installation vessels, turbines to install, prices for vessels or fuel, or the base port's storage capacity. Changes to these parameters reflect structural changes, e.g., the expansion of storage spaces or the commissioning or decommissioning of installation vessels during the project. The last class summarizes the current state of the installation project and, e.g., denotes the current cargo loaded into vessels or their positions. Table 5.1 summarizes the parameters included in the state model. In this table, indexed parameters denote vectors containing one value for each vessel (v), component (c), or instance within the prediction horizon (k). As can be seen from the table, the state comprises some redundant information, e.g., the location of ports and the duration of travel operations. This redundancy results from the fact that the state contains all relevant information for all three optimization models presented in this chapter. These models require information on different abstraction levels. For example, transport optimization requires detailed locations for the production and base ports. In contrast, the scheduling model only requires the resulting traveling times. Consequently, the state update obtains the more detailed parameters (locations) from the real-world system and calculates dependent parameters (traveling times) during the update.

Besides, Table 5.1 shows several parameters that have no symbol assigned. The methodology uses most of these parameters during preprocessing, e.g., to calculate traveling times or operation durations but not directly inside the optimization models. Some other parameters, not having mathematical symbols, act as initial values for various decision variables, e.g., the current vessel location or the current base port inventory.

The methodology calculates two main types of dependent values during the state update before progressing to the estimation of durations for all operations in the next step of the state measurement: First, the update calculates the duration of traveling

Table 5.1 Parameters included in the methodology's state

Symbol	Type	Description
		Optimizer Parameters
T	Time Units	Sampling step width
P	–	Number of planning periods
N	Time Units	Prediction horizon
		Scenario Parameters-Scheduling
N^{owt}	–	Number of turbines to install
V	–	Number of installation vessels
–	date	Project start date
–	hours	Operation base duration
–	m/s	Operation limits – Windspeed
–	m	Operation limits – Waveheight
–	–	Operation sequences – Movement, Installation, Loading
$BAYS$	–	Number of loading bays at the base port
CAP^{port}	Comp. Sets	Maximum port capacity to store component sets
CAP_v	Comp. Sets	Maximum capacity of vessel v to store component sets
RES^{frq}	hours	Minimum time required for base port a resupply cycle
RES^{amt}	Comp. Sets	Number of comp. sets delivered in one resupply cycle
$RES^{initial}$	Comp. Sets	Initial storage of components at the base port
$C_v^{offshore}$	€/hour	Cost for vessel v for being offshore
$C_v^{loading}$	€/hour	Cost for vessel v for loading operations
C_v^{fuel}	€/hour	Fuel cost for vessel v for movement
B^{owt}	(€)	Optimizer benefit for finishing an installation
B^{early}	(€)	Optimizer benefit for finishing an operation early
B^{inv}	(€)	Optimizer penalty for high inventory levels
		Scenario Parameters-Transport
–	long./lat.	Geolocation of ports
–	knots	Traveling speed of HLVs
$CAP^{payload}$	tons	Maximum payload for HLVs
CAP^{area}	m^2	Maximum deck area for HLVs
$REQ_c^{payload}$	tons	Weight of component c
REQ_c^{area}	m^2	Required deck area of component c

(continued)

Table 5.1 (continued)

Symbol	Type	Description
D_c^{setup}	hours	Setup cost for component c on HLVs
D_c^{load}	hours	Loading cost for component c on HLVs
D_c^{unload}	hours	Unloading cost for component c on HLVs
	State Parameters	
–	date	Current date at the start of the sampling step
–	–	Initial location of vessels
–	Comp. Sets	Initial cargo of vessels
–	Comp. Sets	Initial base port inventory
CUR^{owt}	–	Number of turbines finished
$PLAN_v^{vessel}$	date	Earliest instance when a vessel v becomes available
$PLAN_k^{bays}$	–	Number of busy loading bays for each instance k
$PLAN_k^{load}$	–	Number of scheduled loading operations at instance k
$RES^{earliest}$	hours	Minimum number of hours until the base port can receive another shipment of components

times between the respective ports or the installation site using the Haversine formula as shortly summarizes in Sect. 5.3.2.1. Second, it adapts provided forecasts if they do not already include their associated uncertainty, as described in Sect. 5.3.2.2.

5.3.2.1 Calculation of Traveling Times and Distances

The state update calculates all traveling times by dividing the distance between the respective locations by the vessels' speeds, rounding the results to the next full hour. While this approach approximates the traveling time quite well, it relies on estimating the actual traveling distance. In practice, vessels need to follow pre-defined seaways between the involved ports. In contrast, this thesis assumes that vessels travel in a straight line between these locations to increase the methodology's flexibility and achieve a higher degree of comparability with other models and approaches. Most of the approaches found in the literature either use predefined distances or rely on the geographical distance between locations. Other approaches use a GIS to simulate the movement of vessels, e.g., in Ait Alla et al. (2017) or Oelker et al. (2018), which usually apply straight-line routing for sea-based traffic unless the model includes custom made paths for the vessels. This straight-line assumption increases the approach's flexibility during the planning stage, as planners can easily experiment with different combinations of locations without the need to determine the length of actual seaways beforehand. Moreover, it can be assumed that the approximation error in travel times between the actual seaway and the straight line might be neglectable due to the applied rounding. Nevertheless, this straight-line assumption can easily be replaced by a lookup table, stating the respective seaway's length in practical applications.

The proposed methodology uses the Haversine formula as a well-established approximation to calculate the so-called great-circle distance between two geolocations, given by their longitude and latitude. The resulting distance accounts for the earth's curvature, which renders the results comparable to GIS-based simulation models. Given the latitude lat_A, lat_B and longitude lon_A, lon_B of two locations A and B, and the earth radius r, the Haversine formula calculates the distance between these points as:

$$dist = 2r \cdot arcsin$$
$$\left(\sqrt{sin^2\left(\frac{lat_B - lat_A}{2}\right) + cos(lat_A) \cdot cos(lat_B) \cdot sin^2\left(\frac{lon_B - lon_A}{2}\right)}\right)$$

5.3.2.2 Modeling of Forecast Uncertainties

The proposed methodology uses probability intervals for forecasted weather conditions to consider the associated forecasting uncertainty. If forecasts already provide these intervals, the methodology can use them directly. In cases where forecasts only provide the expected value for the wind speed $\mu_s(k)$ and the significant wave height $\mu_h(k)$, the state update calculates these intervals using an estimation of the forecast uncertainty $\delta(k)$. This function depends on the prediction period, i.e., how many hours k the forecast ranges into the future.

This thesis uses the information provided on the Deutscher Wetterdienst's homepage (Deutscher Wetterdienst, 2020) to estimate the uncertainty. The homepage provides an overview of the development of the forecast quality over the last decades and allows the deduction of current forecasting errors depending on the selected forecasting period. The function $\delta(k) \in [0 \ldots 1]$ denotes the uncertainty as a normalized percentage depending on the forecasting period. Based on the information provided by the Deutscher Wetterdienst and discussions with other scientists in the offshore area, $\delta(k)$ interpolates between the following base points: It assumes the uncertainty as 0.0 at the first hour. At 168 hours (one week) into the future, the uncertainty increases to 0.25. At 336 hours (two weeks), the prediction uncertainty amounts to 0.65 and finally increases to 0.95 at 504 hours, from where it quickly increases further. Figure 5.2a sketches the estimated uncertainty related to the forecasting period.

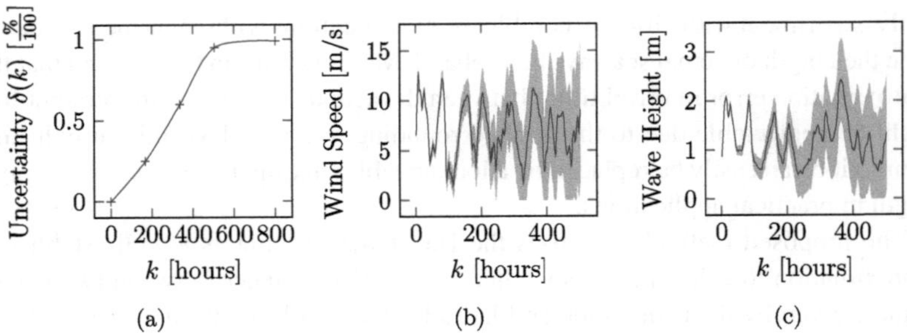

Fig. 5.2 Uncertainty estimation using data from Deutscher Wetterdienst (2020) (a) and probability bands for weather forecasts (area) based on weather measurements (line) for June 2000 (b/c)

When forecasts need a calculation of the probability intervals, the state update calculates the average wind speed μ_s^* and wave height μ_h^* over the prediction horizon N as a reference value. For each hour $k \in N$ in this prediction horizon, it calculates the lower and upper bounds of the probability interval as $forecast_s(k) = \mu_s(k) \pm (\mu_s^* \cdot \delta(k))$ and $forecast_h(k) = \mu_h(k) \pm (\mu_h^* \cdot \delta(k))$ with $\mu_s(k)$ and $\mu_h(k)$ representing the expected value at instance k provided by the forecast. As a result, this method obtains the interval for probable weather conditions, broadening with an increasing prediction period. The interval also regards the average wind speed and wave height within the prediction horizon N. Thus, periods with good weather will also show lower uncertainties, while periods with bad weather show slightly broader probability bands. Figures 5.2b and 5.2c show two exemplary probability bands derived from weather measurements from June 2000. The line depicts the measured or expected value, while the gray area shows the probability band used in the duration estimation.

As noted above, the generation of these intervals or bands allows direct integration of forecasting uncertainties into the methodology. The use of such intervals provides several advantages. For example, if forecasts directly provide estimations on their (time-dependent) quality, this way to adapt the forecasts allows direct integration of this information. If the forecasts do not provide this information directly, the methodology can either rely on the uncertainty model used in this thesis or devise a distinctive model, e.g., relating to the provided data or adapted to practical experiences. The high versatility of this method provides another advantage. On the one hand, the methodology can apply this method to augment forecasts and encode information on their uncertainty. On the other hand, the real-world system's simulation can use this method to generate forecasts from historical weather measurements. As the optimizer only accesses the probability intervals and not the baseline expected value, this method provides an option to "mask" the measurements while providing realistic, measurement-based forecasts for a given period.

5.3.3 Duration Estimation for Offshore Operations

As given in the process description at the beginning of this thesis, each offshore operation requires an uninterrupted time window, which provides suitable weather conditions to finish. If the conditions would exceed the operation's limits at any time, it would need to abort and completely restart again once the weather conditions drop below the given limits again. To avoid such situations, this thesis considers the total duration of an operation as its processing time plus the necessary waiting time to reach a suitable weather window. The methodology first estimates each basic operation's total duration at each starting hour k, using the weather restrictions and the provided forecasts. Afterward, it summarizes these durations to calculate the duration of each aggregate operation o as $D_{o,k}$, by chaining the respective durations.

Figure 5.3 shows an example of this aggregation. This example shows that the aggregate operation consists of the basic operations *Reposition*, *Jack-up*, and *Install Tower*, and other operations not listed in this figure. The aggregate operation starts at the first time instance $k = 0$. As the *Reposition* operation, starting at $k = 0$ takes two hours, the algorithm next acquires the duration of the *Jack-up* operation, assuming that it starts immediately after the repositioning finishes, i.e., at instance $k = 2$. This operation again takes two hours, resulting in a starting time of the *Install Tower* operation at $k = 4$, where it takes three hours to finish. By traversing all basic operations within an aggregate operation, the algorithm calculates the aggregate one's total operation time.

Hour	0	1	2	3	4	5	6	7	...
Reposition	2	1	1	1	1	3	2	1	...
Jack-up	4	3	2	2	2	4	3	2	...
Tower	4	3	3	3	3	4	3	3	...
...	...	...	...	...	...	...	...	...	...

Fig. 5.3 Example of the calculation of aggregate operations' durations for starting hour $k = 0$—derived from Rippel et al. (2019d) © 2019 IEEE

This process relies on an estimation of the basic operations' durations. While the methodology can easily calculate these values as a combination of processing and waiting times for known or deterministic weather conditions, the integration of weather forecast uncertainties requires specialized methods to achieve a good estimation. Figure 5.4 shows an example of the simplest case of known weather conditions. In this example, the methodology simply traverses the time series to identify the first instance of a sufficiently large weather window of at least three hours with wind speeds lower than 12 m/s and wave heights lower than 2.5 m. It then considers the combination of waiting (6 hours) and processing times (3 hours) as total duration at instance $k = 0$. This simple case only occurs during the process simulation, where the simulation component accesses historical records to verify the methodology's plans against real weather conditions.

	Waiting Time						Processing Time			
Hour k	0	1	2	3	4	5	6	7	8	9
Wind Speed	13.0	12.8	12.4	12.0	11.9	12.1	11.6	11.5	11.1	10.8
Wave Height	3.1	2.9	2.7	2.5	2.4	2.3	2.1	1.8	1.8	1.5

Start: $k = 0$ Duration: 3 hours

Max Wind: 12 m/s Max Wave: 2.5 m

Fig. 5.4 Example of calculating operation durations for known weather conditions

The scheduling methodology itself only accesses forecasts, consisting of intervals for probable wave heights and wind speeds at each instance k. The encoded forecast uncertainty and possible errors within the forecasts require specialized methods to estimate each operation's duration accurately. Therefore, this thesis proposes two possible methods. The first method uses sliding windows, similar to the case described above. However, it interprets the forecast intervals in terms of probability distributions to select windows with a sufficiently high probability of finishing the operation. The second method uses discrete-time Markov Chains to estimate when an aggregate operation could finish. The following sections shortly describe both estimators as examples of how parts of the overall methodology could be exchanged. The experiments described later in this thesis generally use the Markov Chain estimator, as it shows slightly better results. Please refer to Sect. 8.1.1 for a comparison of the runtime-behavior of both estimators.

5.3.3.1 Estimation by Sliding Windows

The sliding window estimator follows the same general idea as before. When starting an operation at a particular hour, this estimator compares the limits for this operation with the respective forecasts to find suitable weather windows. It comprises two steps. First, it calculates the probability of successfully conducting the operation at each possible instance within the prediction horizon. Second, it applies the name-giving sliding window to determine a suitable time window.

As the forecasts provide an interval of probable wind speeds and wave heights, this estimator first calculates the probability of conducting the operation successfully at the given instance. The current implementation interprets the forecast interval as a uniform distribution to simplify computations and handle the most general case of not having any additional information on the forecasts' internal structure. The assumption of a uniform distribution allows calculating the probability of successfully conducting the operation as the fraction of the interval, which is lower or equal to its limit. Figure 5.5a shows an example. While this assumption shows acceptable results, the sliding window estimator considers and implements the probability calculation as s distinctive step. This separation allows exchanging the underlying assumptions, e.g., to include more complex calculations or adapt this assumption if more information about the forecasts is available. Such adaptions might include the expected trajectory of the development of wave heights or wind speeds or use advanced assumptions about the distribution if known for the used forecast data.

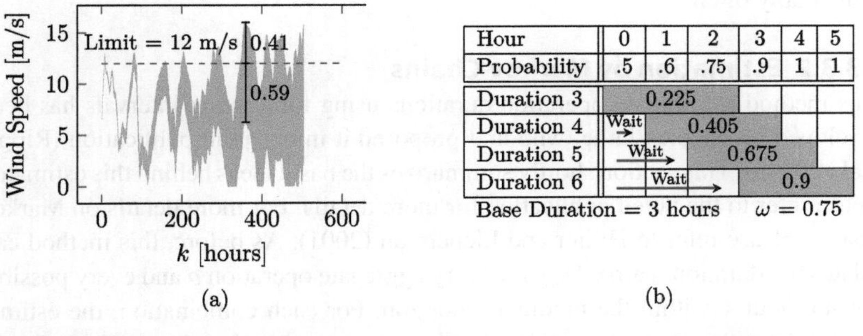

(a) (b)

Fig. 5.5 (a) Sliding window estimation: calculation of probabilities and (b) application of the sliding window—derived from Rippel et al. (2019d) © 2019 IEEE

The second step finally estimates each operation's total duration for each starting hour within the prediction horizon to construct the matrix $D_{o,k}$. For each hour, this estimator first selects the lower probability, either for the wind speed or the wave height, to ensure that both weather conditions satisfy the operation's restrictions. Figure 5.3 shows an application example of the sliding window method. First, it constructs the sliding window with a width according to the operation's base duration at the selected starting hour (first row—duration 3). The method then calculates the joint probability for successfully finishing the operation within this window and compares it with a user-defined threshold ω. If the joint probability exceeds this threshold, the method accepts the window and returns the total duration, again given by the sum of waiting and processing times. If the joint probability lies below the threshold ω, the estimator rejects the window and moves the window one hour ahead, thus, expecting the vessel to wait for one additional hour. The method repeats this process until it detects a suitable window (second to fourth row).

This estimator provides users with the opportunity to modify the estimation's robustness by selecting the threshold ω. If the process planner selects a high ω, the resulting estimations move towards a robust, worst-case assumption as the estimator requires a high joint probability to accept a window. When selecting $\omega = 1.0$, the estimator only accepts windows that certainly allow the operation to finish, drawing the overall methodology close to a robust approach, as introduced in the state of the art. If process planners choose a low ω, the selected time windows will include an increasing number of improbable windows, effectively resulting in a high number of recourse actions as the estimations might turn out increasingly wrong. As with a high ω, extremely low values for ω reduce the predictive capabilities of this approach as the estimator will accept close to any time window and, thus, plans will fail comparably often.

5.3.3.2 Estimation by Markov Chains

This method to estimate operation durations using forecasting intervals has been developed by Nicolas Jathe, who first proposed it in our joint publication (Rippel et al., 2019b). This section shortly summarizes the basic ideas behind this estimator. Please refer to the noted publication for more details. For more details on Markov Chains, please refer to Hillier and Lieberman (2001). As before, this method calculates the duration matrix $D_{o,k}$ for every aggregate operation o and every possible starting hour k within the prediction horizon. For each combination, the estimator simulates the operation as a stochastic process, measuring the number of steps required to finish the operation with a given probability.

To simulate the stochastic process, the estimator first constructs a discrete-time Markov Chain. Each state represents one hour of progress, with a final state added to represent finishing the aggregate operation. This Markov Chain distinguishes between normal states and helper states. Each normal state either transitions to itself or its follow-up state. This construct represents that the operation has either started or needs to wait due to insufficient weather conditions. Helper states model the progression of time for operations and only progress to the next state with a probability of 1.0.

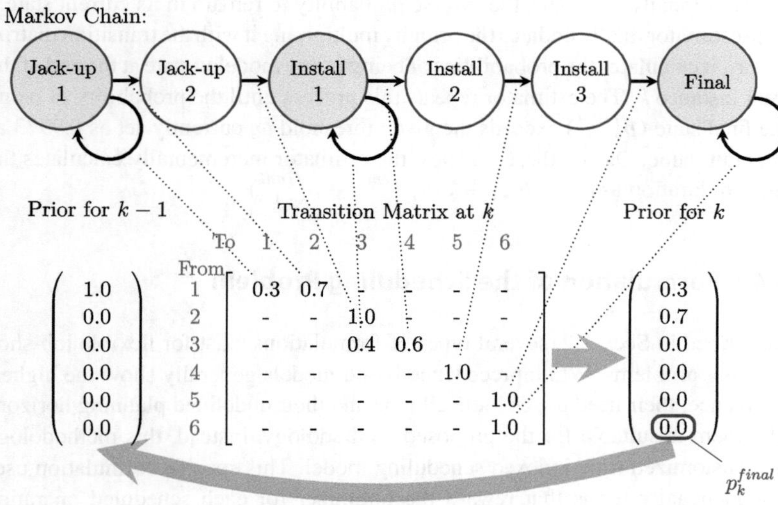

Fig. 5.6 Internal structure of the duration estimation using discrete-time Markov Chains

Figure 5.6 shows an exemplary Markov Chain at the top, depicting an aggregate operation consisting of the two basic operations *Jack-up* and *Install Tower*, with durations of two and three hours. Consequently, the Markov Chain consists of six states—one normal and one helper state for the jack-up operation, one normal and two helper states for the installation, and one as the final, absorbing state. Next, the estimator denotes the probability of being in each state at the current time k as vector or so-called prior. Figure 5.6 shows two priors for the instance $k-1$ and k at its left and right. Initially, this prior assigns a probability of one to the first state and zero for all others. The estimator then simulates the progression of the aggregate operation.

First, it obtains the weather and operation dependent probabilities of transitioning out of each normal state for the current instance k. Like the first estimator, this one compares the weather restrictions of the state's basic operation to the current weather forecasts. In contrast, it assumes that the forecast interval represents a three-sigma normal distribution, calculating the probability of suitable weather accordingly.

Moreover, it chooses the minimum probability over the required weather window, i.e., the minimum over the next two or, respectively, three hours in the example given above. The estimator then constructs the Markov Chain's transition matrix by assigning the calculated probability as the probability to transition into the next state. Additionally, it assigns the inverse probability to remain in its current state.

The estimator finally updates the prior by multiplying it with the transition matrix. This way, it calculates the probabilities of being in any modeled state at the end of the current instance k. The estimator repeats this process until the probability of being in the final state (p_k^{final}) exceeds the given threshold ω, currently set as 0.9973 as the default value. During these updates, the estimator incrementally calculates the estimated duration as $d_k = d_{k-1} + k \cdot (p_k^{final} - p_{k-1}^{final})$.

5.3.4 Formulation of the Scheduling Problem

As described in Sect. 3.2, several types of formulations exist for flexible job-shop scheduling problems. While precedence-based models generally show the highest performance, their need to schedule all jobs and their undefined planning horizons render them unsuitable for the proposed methodology. Instead, this methodology uses a customized time-indexed scheduling model. This specific formulation uses so-called penalty terms that reward the optimizer for each scheduled operation instead of requiring it to schedule all operations. This formulation allows adapting each instance of the scheduling problem to the desired prediction horizon N, depending on the quality of weather forecasts and computational limits.

The proposed formulation tries to schedule sequences of loading, movement, and installation operations to minimize the costs for operations and offshore waiting times. Moreover, this model includes port-side restrictions regarding available loading bays or cargo bridges and the resupply of turbine components. As described above, the model relies on a discretized estimation for the duration of operations to avoid the need for expensive stochastic optimization techniques.

5.3.4.1 Variables and Parameters

The proposed model relies on several decision variables and includes other variables, which directly depend on these decision variables. The optimizer assigns sequences of operations that minimize installation's overall cost by using the cost function and the constraints described below. The overall problem structure resembles a mostly binary matrix, which denotes time instance (hours) on the horizontal axis and the operations on the vertical axis, as shown exemplarily in Table 5.2. The formulation uses a total of five decision variables, one for each type of operation: installation

$Y_{v,k}^{owt}$, move to port $Y_{v,k}^{toPort}$, move to site $Y_{v,k}^{toSite}$, and loading $Y_{v,k}^{loading}$, plus one decision variable that allows to schedule resupply operations that increase base port's component inventory $Y_{k}^{resupply}$. In addition, the model uses a total of seven dependent variables to track the location $Y_{v,k}^{loc}$, and busy state of vessels $Y_{v,k}^{busy}$, loading bay occupancies $Y_{v,k}^{atBay}$, and inventory levels for the port X_{k}^{invP} and the vessels $X_{v,k}^{invV}$. Moreover, the model uses two additional dependent variables to track the finishing times of installation operations $X_{v,k}^{fOwt}$, and other minor operations $X_{v,k}^{fOp}$. The cost function uses these variables to assign bonuses for finishing operations early within the prediction horizon.

Table 5.2 shows a minimal example of these variables for a single vessel. It can be seen that the vessel starts a loading operation at the first hour of the resulting plan, finishes an operation at N-5, and is idle at hour N-4. It then moves back to the port at the end of the plan (N-3), which takes a total of four hours.

Table 5.2 Example for decision variables of a single vessel

$v = 1$	1	2	3	…	N-5	N-4	N-3	N-2	N-1	N
$Y_{v,k}^{owt}$	0	0	0	…	0	0	0	0	0	0
$Y_{v,k}^{toPort}$	0	0	0	…	0	0	1	0	0	0
$Y_{v,k}^{toSite}$	0	0	0	…	0	0	0	0	0	0
$Y_{v,k}^{loading}$	1	0	0	…	0	0	0	0	0	0
$Y_{v,k}^{busy}$	0	1	1	…	1	0	0	1	1	1
…	…	…	…	…	…	…	…	…	…	…

Due to the integration with the MPC, the model requires information about previous planning cycles, especially if scheduling several installation vessels. In such cases, each vessel can finish at different times, e.g., the first vessel's plan covers 100 hours while the second vessel needs to abort plan execution at hour 80. In this case, a new iteration would start at hour 80, but the first vessel still has part of the previous plan left. To avoid unnecessary rescheduling of these remaining plans, the model blocks vessels until their plans have finished using the parameter $PLAN_{v}^{vessel}$, which denotes the earliest date/hour the vessel can start. This parameter can also describe vessels, which become available later during the project e.g. if the project starts with a single vessel in April but adds a second one in May. Using similar parameters, the model tracks scheduled loading operations ($PLAN_{k}^{load}$) to manage inventories across plans and bay occupancies ($PLAN_{k}^{bays}$). The tracking of used loading bays/bridges results from the model's ability to limit port side resource availabilities. As ports still conduct normal operations, it can be expected that only a limited number of loading bays or cargo bridges have been reserved for the installation project. The model represents this amount using the parameter $BAYS$, which effectively limits the number of simultaneous loading operations for installation vessels.

The model tracks the inventory and resupply of components to the base port and on the installation vessels. Thereby, the parameters CAP^{port} and CAP_v describe the maximum inventory levels at the base port and each vessel. Considering the resupply of components, the model allows the optimizer to schedule a specific operation type, so-called resupply operations. In practice, one or more HLVs follow a predefined resupply cycle, fetching components from their production ports and transporting them to the base port for further processing. The proposed model simplifies these cycles by only considering full sets of components. Thus, the model describes these resupply cycles using their duration RES^{frq} and the number of complete sets to deliver RES^{amt}. Thus, every RES^{frq} hours, the base port's inventory would increase by RES^{amt}. To avoid situations where the delivered amount would exceed the maximum inventory CAP^{port}, the model can postpone these deliveries by scheduling the mentioned resupply operations. Nevertheless, it can only schedule additional resupply operations after RES^{frq} hours have passed since the last operation. To track the remaining waiting time, the model uses the parameter $RES^{earliest}$.

5.3.4.2 Cost Function

The optimization problem minimizes the cost function J given in equations (5.1) – (5.3). The cost function consists of two parts. The first part in equation (5.2) calculates the actual costs of the scheduled operations by summing the costs for the time each vessel spends offshore ($C_v^{offshore}$) and the costs for loading ($C_v^{loading}$) and movement operations (C_v^{fuel}). The second part given in equation (5.3) consists of several penalty terms.

$$J(x, u) = J^c(x, u) + J^p(x, u) \tag{5.1}$$

$$\begin{aligned} J^c := \sum_{v=1}^{V} \Bigg(\sum_{k=1}^{N} \left(Y_{v,k}^{loc} \right) \cdot C_v^{offshore} \quad & + \sum_{k=1}^{N} \left(Y_{v,k}^{atBay} \right) \cdot C_v^{loading} \\ + \sum_{k=1}^{N} \left(Y_{v,k}^{toSite} \right) \cdot C_v^{fuel} \quad & + \sum_{k=1}^{N} \left(Y_{v,k}^{toPort} \right) \cdot C_v^{fuel} \Bigg) \end{aligned} \tag{5.2}$$

$$J^p := \left(\sum_{k=1}^{N} X_k^{invP}\right) \cdot \frac{1}{N \cdot CAP^{port}} \cdot B^{inv} \qquad - \sum_{v=1}^{V} \left(\sum_{k=1}^{N} \left(Y_{v,k}^{owt} \cdot B^{owt}\right)\right.$$
$$- \sum_{k=1}^{N} \left(\frac{X_{v,k}^{fOwt}}{N} - 1\right) \cdot -B^{early} \qquad - \sum_{k=1}^{N} \left(\frac{X_{v,k}^{fOp}}{N} - 1\right) \cdot \left(- 0.2 \cdot B^{early}\right)\Bigg) \tag{5.3}$$

5.3.4.3 Description and Justification of Cost Terms

The cost function focuses on minimizing the variable costs of scheduled operations, i.e., installation, loading, and movement operations, as given in equation (5.2). Consequently, it omits fixed costs, e.g., for renting vessels over a period of time. As stated in the process description, vessels come at daily charter rates between 70,000€ and 330.000€. While it would be possible to include these costs in the optimization model directly, the model only covers the additional hourly costs that arise from the vessel being offshore as $C_v^{offshore}$. These costs include, e.g., additional contractual costs, personnel costs, and a baseline amount of fuel used to power the vessel's basic systems. As give in Sect. 2.2.4, these costs amount to approximately 20% of the daily charter rates.

This setup assumes that project managers need to charter vessels for a fixed amount of time in advance and, thus, these charter rates can be accounted to the fixed costs. It has to be noted that contracts usually include paragraphs on shortening or extending the charter period so that a faster installation will reduce the overall charter costs. On the one hand, it would be possible to enforce shorter plans by extending the formulation to minimize the makespan directly or by including charter costs over the final plan duration. On the other hand, the penalty terms introduced later in this section already provide a bonus to the optimizer, the earlier it schedules an operation. While both options would be viable, the consideration of the makespan would introduce another cost term that would most probably outweigh the other costs by far. In consequence, the costs would require some kind of artificial scaling to prevent the optimizer from primarily generating as short as possible plans, mainly ignoring variable operating costs. The penalty terms introduced later also facilitate the generation of shorter plans. This way of formulation has two advantages. First, it favors starting operations early if there would be no difference in their duration. Second, it favors shorter operations over ones that come with waiting times. Moreover, this formulation allows a stronger focus on minimizing variable costs compared to minimizing the overall makespan.

The second type of cost covered by the cost function constitutes additional costs for renting equipment at the base port for loading operations, denoted as $C_v^{loading}$.

While general costs for storage and port operations most likely account as fixed costs covered in long-term contracts, most ports offer the rental of heavy-duty equipment on an hourly basis (see, e.g., EUROGATE (2019) for an example). Consequently, each port operation adds a specific cost to the overall variable costs, depending on its duration. It has to be noted that this cost term also serves a technical purpose: It issues bonuses for each conducted operation. Thus, each scheduled operation should induce additional costs to prevent the optimizer from planning as many operations as possible to get these bonuses.

While the costs for loading operations generally remain static due to the port's weather protected environment, movement operations may vary in duration. Moreover, installation vessels require large amounts of fuel when traveling compared to the amount consumed in standby mode. Some shipping companies provide spec-sheets of installation vessels on their homepage, e.g., Fred. Olsen Windcarrier (2020), which show drastic differences in the consumption if the vessels move or remain stationary, easily covering about 30 to 40 tons a day.

5.3.4.4 Description and Justification of Penalty Terms

The penalty terms in equation (5.3) alter the optimizer's behavior, e.g., by assigning a bonus, i.e., negative costs, according to B^{owt}, for each finished OWTs (second term). As the overall problem constitutes a minimization problem and all operations incite costs, the optimizer would, by itself, not schedule any operation. Thus, the formulation needs to either enforce the assignment of all, or at least a number of operations by constraints or provide additional incentives, like this penalty term. While standard formulations found in the literature generally resort to the first solution, the approach proposed in this thesis assumes that the prediction horizon will, in most cases, be too short to fit all operations. Thus, adding such a constraint would require choosing unrealistically long prediction horizons or render the optimization problem infeasible.

The first term of equation (5.3) directs the optimizer to either keep the inventory levels of components high when choosing negative values for B^{inv} (default) or as low as possible. The formulation presented below allows the optimizer to schedule resupply operations whenever a resupply cycle, described by its duration RES^{frq} and the amount of component sets RES^{amt} finishes. These operations add the corresponding amount of sets to the base ports storage but can only be scheduled if the storage has enough free space and when a minimal waiting time since the last delivery has passed. Thus, the optimizer shows the best performance when it keeps inventory levels as high as possible, and schedules resupply operations as soon as they become available. Nevertheless, the parameter B^{inv} can instruct the optimizer to schedule these operations only when needed, e.g., to evaluate just-in-time strategies.

The last two terms aim to reduce the problem formulation's symmetry by assigning increasing benefits the earlier an operation finishes. Thus, if an installation operation would finish in the first hour of the prediction horizon, these terms would issue an additional bonus of B^{early}, which linearly decreases for operations finishing in the last hour of the prediction horizon. The last term assigns a similar bonus for non-installation operations but scales these by 0.2 as a randomly picked scaling factor. This scaling prevents the optimizer from scheduling minor operations just to obtain the associated bonus. As noted, these terms try to reduce the symmetry of the optimization problem by avoiding equal costs for dissimilar solutions. For example, imagine a situation with perfect weather and a prediction horizon that could fit one installation operation, which could start at hour one, two, or three and finish with the same duration. This example represents a symmetrical situation where all three options result in the same costs, rendering the optimization problem ill-posed. By favoring the solution finishing earlier (hour one) over the others, the presented cost function removes this symmetry, allowing the optimizer to solve the problem more efficiently.

While the model tracks each vessel's offshore times and times at loading bays by the hour, it only denotes the occurrence of movement operations. Therefore, the algorithm scales the costs for these operations by their base duration before calculating the cost function. While this scaling introduces an approximation and neglects extended operation times due to weather influences, these particular operations have comparably high limits. Thus, cases, where weather prolongs the execution time occur very rarely. If required, the model can easily be extended to track the actual duration similar to offshore times, at the cost of additional dependent variables and constraints.

5.3.4.5 Constraints

Besides the cost function, the proposed formulation uses several constraints given in equations (5.4) – (5.23). This subsection introduces these constraints and shortly describes their structure and the reasoning behind them. Several constraints thereby depend on the last time instance $k - 1$. As given below, these usually only apply for all k in $\{2 \dots N\}$. While not explicitly listed here, each of these constraints has a counterpart for $k = 1$ that retrieves these last values from the provided state for the current planning instance $x(t)$ as initial values.

Constraint (5.4) tracks the current amount of loaded components for each vessel. This value increases when the vessel loads a set of components at the base port and decreases for every started installation operation. Thereby, constraint (5.5) enforces that the vessels' inventories never exceed the maximum capacity given as CAP_v.

$$X^{invV}_{v,k} = X^{invV}_{v,k-1} - Y^{owt}_{v,k} + Y^{loading}_{v,k} \qquad \forall k, v \quad 2 \leq k \leq N, v \leq V \tag{5.4}$$

$$X^{invV}_{v,k} \leq CAP_v \qquad \forall k, v \quad k \leq N, v \leq V \tag{5.5}$$

Similarly, constraint (5.6) tracks the location of the vessel. This binary variable can either take the value 1 (offshore) or 0 (at the base port). Thus, this variable increases when a vessel moves to the site and decreases when it moves to the port. The constraint (5.7) enforces that the vessel needs to be offshore to start installing an OWT. Equally, constraint (5.8) requires the vessel to be in the port to load components.

$$Y^{loc}_{v,k} = Y^{loc}_{v,k-1} - Y^{toPort}_{v,k} + Y^{toSite}_{v,k} \qquad \forall k, v \quad 2 \leq k \leq N, v \leq V \tag{5.6}$$

$$Y^{loc}_{v,k} \geq Y^{owt}_{v,k} \qquad \forall k, v \quad k \leq N, v \leq V \tag{5.7}$$

$$Y^{loc}_{v,k} \leq (1 - Y^{loading}_{v,k}) \qquad \forall k, v \quad k \leq N, v \leq V \tag{5.8}$$

While the former constraints manage the start of movement, loading, and installation operations, the constraints (5.9) – (5.12) track the duration of operations and mark them as part of the dependent variable $Y^{busy}_{v,k}$. Constraint (5.14) relies on these busy times to ensure that a single vessel only performs a single action at once. These constraints use the matrix of operation durations $D_{o,k}$ calculated earlier and provided as part of the state to track busy times. Constraint (5.13) tracks the duration of loading operations, in the same way, using the dependent variable $Y^{atBay}_{v,t}$. Later on, constraint (5.19) uses this variable to ensure that vessels cannot use more loading bays than available.

$$\sum_{t=k+1}^{k-1+D_{1,k}} Y^{busy}_{v,t} \geq Y^{owt}_{v,k} \cdot (D_{1,k} - 1) \qquad \forall k, v \quad k \leq N, v \leq V \tag{5.9}$$

$$\sum_{t=k+1}^{k-1+D_{2,k}} Y^{busy}_{v,t} \geq Y^{toPort}_{v,k} \cdot (D_{2,k} - 1) \qquad \forall k, v \quad k \leq N, v \leq V \tag{5.10}$$

$$\sum_{t=k+1}^{k-1+D_{3,k}} Y_{v,t}^{busy} \geq Y_{v,k}^{toSite} \cdot (D_{3,k} - 1) \qquad \forall k, v \quad k \leq N, v \leq V \tag{5.11}$$

$$\sum_{t=k+1}^{k-1+D_{4,k}} Y_{v,t}^{busy} \geq Y_{v,k}^{loading} \cdot (D_{4,k} - 1) \qquad \forall k, v \quad k \leq N, v \leq V \tag{5.12}$$

$$\sum_{t=k}^{k-1+D_{4,k}} Y_{v,t}^{atBay} \geq Y_{v,k}^{loading} \cdot D_{4,k} \qquad \forall k, v \quad k \leq N, v \leq V \tag{5.13}$$

$$Y_{v,k}^{owt} + Y_{v,k}^{loading} + Y_{v,k}^{toPort} + Y_{v,k}^{toSite} + Y_{v,k}^{busy} \leq 1 \quad \forall k, v \quad k \leq N, v \leq V \tag{5.14}$$

Constraint (5.15) ensures that the optimizer only schedules installation operations that finish within the prediction horizon. Constraint (5.16) ensures that the optimizer does not schedule the installation of more OWTs than necessary.

$$Y_{v,k}^{owt} \leq max(0, (N - k) - (D_{1,k} - 1)) \qquad \forall k, v \quad k \leq N, v \leq V \tag{5.15}$$

$$\sum_{k=1}^{N} \sum_{v=1}^{V} X_{v,k}^{owt} \leq N^{owt} - CUR_{t_i}^{owt} \tag{5.16}$$

Constraints (5.17) and (5.18) record the expected finish times of scheduled operations. The cost function uses these values to assign a bonus for operations that finish early within the prediction horizon to reduce the model's symmetry.

$$X_{v,k}^{fOwt} = Y_{v,k}^{owt} \cdot (k + D_{1,k} - 1) \qquad \forall k, v \quad k \leq N, v \leq V \tag{5.17}$$

$$\begin{aligned} X_{v,k}^{fOp} &= Y_{v,k}^{toPort} \cdot (k + D_{2,k} - 1) \\ &+ Y_{v,k}^{toSite} \cdot (k + D_{3,k} - 1) \\ &+ Y_{v,k}^{loading} \cdot (k + D_{4,k} - 1) \qquad \forall k, v \quad k \leq N, v \leq V \end{aligned} \tag{5.18}$$

As described earlier, the proposed model needs to prevent a rescheduling of vessels or loading bays if these are already in use by another plan from the previous scheduling iteration. Therefore, constraint (5.19) enforces that the number of currently used bays remains lower or equal to the available bays. It uses the parameter $PLAN_k^{bays}$ to ensure that bays cannot be rescheduled if old, still valid plans block them. Constraint (5.20) uses the same strategy with parameter $PLAN_v^{vessel}$ to ensure that vessels, which still have valid plans from the last schedule, cannot be rescheduled for this time.

$$\sum_{v=1}^{V} Y_{v,k}^{atBay} + PLAN_k^{bays} \leq BAYS \qquad \forall k \;\; k \leq N \tag{5.19}$$

$$Y_{v,k}^{busy} \leq max(0, k - PLAN_v^{vessel}) \qquad \forall k, v \;\; k \leq N, v \leq V \tag{5.20}$$

In terms of the base port's resupply, the following constraints track the current inventory levels at the base port (5.21), the adherence to the minimal duration of resupply circles (5.22), and the carry-over from previous schedules (5.23).

$$X_k^{invP} = X_{k-1}^{invP} - \sum_{v=1}^{V} Y_{v,k}^{loading} + Y_k^{resupply} \cdot RES^{amt} - PLAN_k^{load} \qquad \forall k \;\; 2 \leq k \leq N \tag{5.21}$$

$$\sum_{t=max((k-RES^{frq}),1)}^{k} \left(Y_t^{resupply} \leq 1\right) \qquad \forall k \;\; k \leq N \tag{5.22}$$

$$\sum_{t=1}^{min(N,RES_{t_i}^{earlies})} \left(Y_t^{resupply} = 0\right) \tag{5.23}$$

5.3.5 Control Extraction and Application

After finishing the open-loop optimization, the methodology finally extracts a short-term plan for each scheduled vessel and forwards these plans to the real-world system or the simulation component. Therefore, it truncates every plan at the last operation that finishes before T. Afterward, it records all required parameters for the next sampling instance, assuming that all plans will proceed as planned. First, it schedules the next sampling step at the end of the shortest of these plans. For each longer plan, it denotes how many hours each vessel's plan exceeds this shortest plan to prevent a rescheduling during this period. Moreover, it denotes planned loading operations to track the usage of loading bays and the current inventory level. Finally, it forwards the plans for execution.

As the overall methodology acts as a DSS, it monitors the execution of plans. The steps above suffice to ensure a seamless online optimization of the installation process if everything proceeds as planned. Additionally, the methodology monitors the execution of each operation to track possible plan failures. The methodology expects notifications from the real-world system on how long a planned operation takes to finish in the real-world during the execution. It then compares the planned duration with the operation's real duration. If the durations match or an operation takes longer, but the next one still starts as planned, the methodology considers this operation successful. In cases where a vessel misses a follow-up operation or if an operation takes less time than scheduled, the methodology considers the plan failing. In such a case, the methodology triggers a recourse action, resulting in an early start of the next sampling step. Therefore, it first repeats the steps noted above to synchronize all valid plans with the current time, i.e., it updates the remaining length of valid plans and the parameters denoting planned loading operations. Afterward, it starts a new scheduling cycle by measuring the real-world system's current state.

5.4 Models for Offline Planning Tasks

This section presents two models for planning tasks on tactical or strategic planning levels to evaluate if the proposed scheduling methodology can sufficiently support these tasks. First, this section presents a model for planning transports for the resupply of components to the base port in Sect. 5.4.1. Second, it presents a model for optimizing the base port's capacity, depending on the employed resupply cycles and the installation project's characteristics in Sect. 5.4.2. As project managers usually need to decide on the resupply cycles and capacities long before the actual installation project starts, these models use an offline optimization based on historical data. Nevertheless, in particular, capacity optimization utilizes several of the methods proposed as part of the above methodology.

5.4.1 Transport Planning

While the previously presented model constitutes the main focus of this thesis, the methods and approaches presented as part of the scheduling methodology can support other planning tasks as well. This section presents a model for the transport planning of components between the production ports and the base port.

In practice, project managers usually plan the resupply of components before the actual installation project starts. Therefore, they define a number of round-trips between the base ports and the respective production ports. During the installation, one or more HLVs follow these round-trips and transport a pre-defined number of components to the base port. This thesis refers to such a set of round-trips as a resupply cycle. Thereby, each cycle provides a given amount of component sets RES^{amt} with a predetermined duration RES^{frq} within a set amount of round-trips N^{rt}. Depending on the HLVs capacities in terms of their maximum payload and deck area, each round-trip picks up a number of components from at least one of the production ports before proceeding to the base port for unloading.

5.4.1.1 Assumptions and General Algorithm

This algorithm focuses on the minimization of the resupply cycle's duration while maximizing its yield. These parameters act as input for the scheduling methodology presented above. Therefore, the algorithm assumes a set number of round-trips and tries to optimize the number and type of components picked-up and delivered in each round-trip. Each of these round-trips starts and ends at the base port but can visit any combination of production ports. This problem's general notion constitutes a so-called transshipment problem as, e.g., described by Agadaga and Akpan (2017). These problems aim to find the optimal route to deliver cargo from several storages (sources) to a number of depots (sinks). Thereby, each source can also act as a sink and vice versa. This thesis's particular instance consists of three sources (production ports) and a single sink (base port). Moreover, the involved HLV provides a limited capacity, given by its deck area CAP^{area} and maximum payload $CAP^{payload}$. Accordingly, each component requires a certain space REQ_c^{area} and has a specific weight $REQ_c^{payload}$. Consequently, the described problem combines a standard routing problem to minimize traveling times with a variant of an assignment problem to choose which mix of components to pick in each round-trip j.

The proposed algorithm takes advantage of the transshipment network's small size to reduce the problem's complexity and separate both problems. Therefore it first solves a small number of TSPs to obtain the duration, i.e., costs of all possible routes the HLV could take. Second, the algorithm applies a custom-made formulation for a multi-periodic Knapsack problem to determine the optimal mix of components for each round-trip, employing the pre-calculated route costs. The algorithm takes several assumptions regarding travel times to separate these problems:

- Instead of monetary costs, this algorithm uses the duration of the resupply cycle for its cost function. Nevertheless, these durations could be recalculated to currency easily.
- This algorithm ignores the influence of weather on the resupply of components. This assumption results from the fact that traveling and loading/unloading operations at the ports constitute very robust operations that are only affected by comparably high wind speeds and wave heights. Moreover, the transport planning, as considered in this section, occurs long before actual weather information could be retrieved. Thus, this model assumes a travel time according to the distance divided by the vessel's speed. For safety reasons, project managers could choose to add a security margin to all durations.
- The presented algorithm considers routes to be symmetrical. Thus, traveling from port A to port B incurs the same costs as traveling from port B to port A.
- As with all other algorithms presented in this thesis, this algorithm assumes the distance between two locations as the direct "air-line" distance instead of actual seaways. This thesis calculates all distances based on the longitude and latitude using the Haversine formulae to compensate for the earth's curvature. In practical applications, project managers can replace these distances with actual seaways either from lookup-tables or by including a viable routing solver for seaways.

5.4.1.2 Variables and Cost Function

The optimization model uses the decision variable $X_{j,c}^{freight}$ to optimize how many components $c \in \mathbb{N}^+, c \leq N^c$ the vessel should fetch in each round-trip $j \in \mathbb{N}^+, j \leq N^{rt}$. The model uses a dependent, binary variable $Y_{j,c}^{port}$ to denote which port the vessel visits in each round-trip. For each round-trip j, this variable consists of a binary array $[p_t, p_b . p_n]^T$. For example, the array $[1, 0, 1]^T$ denotes that the vessel visits the production port for towers p_t and nacelles p_n but not the blades' production port p_b. The algorithm then indexes these arrays by using a standard binary encoding to convert them to unique indices. As it considers routes symmetrical, the total number of existing routes that a vessel can take in each round-trip is given as 2^{N^c}, resulting in a total of eight possible combinations of visited ports. Using a binary encoding, as given in equation (5.24), the algorithm assigns a unique index $r \in \mathbb{N}^+, r \leq 2^{N^c}$ to each of these combinations.

$$r = Y_{j,1}^{port} \cdot 2^0 + Y_{j,2}^{port} \cdot 2^1 + Y_{j,3}^{port} \cdot 2^2 \quad \forall j, \quad j \leq R \tag{5.24}$$

The optimization model uses this unique mapping to obtain the precalculated costs for each route. Therefore, the algorithm solves a standard TSP for each possible route and stores each route's duration as vector D_r^{route}. The model uses the binary variable $Y_{j,r}^{route}$ to denote which of the routes r corresponds to the visited production ports for each round-trip j. In addition to these durations, the optimization model also considers loading (D_c^{load}) and unloading (D_c^{unload}) times for each component as well as possible setup times (D_c^{setup}). Depending on the vessel and the involved companies, port operatives might need to mount or unmount specific transport frames for each component. Thus, the model also tracks possible setup operations as the difference in cargo between round-trips using the dependent variable $X_{j,c}^{setup}$. For example, if the first round-trip would transport two towers and one set of blades and the second round-trip consists of one tower and three sets of blades, the algorithm would denote setup operations for one tower ($2 \rightarrow 1$) and one set of blades ($2 \rightarrow 3$).

Figure 5.7 depicts the structure of the optimization problem and the dependencies between its variables. In this theoretical example, the optimizer determined that for a given number of two round-trips, the vessel should pick up four towers and two sets of blades in the first trip, and two sets of blades, and four nacelles in the second. Accordingly, it noted possible setup operations as the difference between each round-trip and its predecessor (center bottom) and the visited production ports

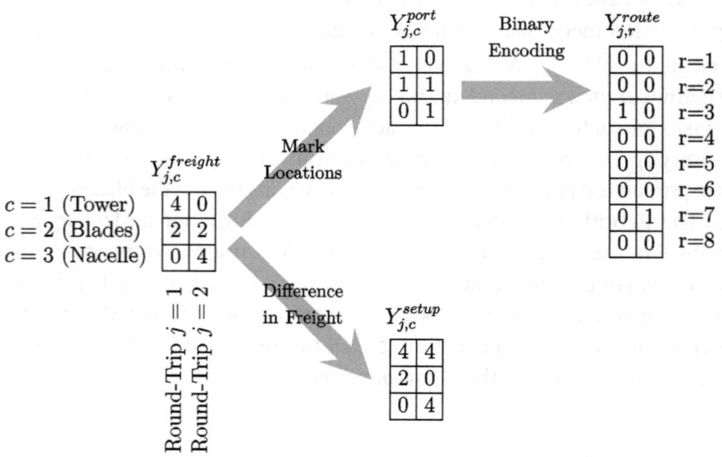

Fig. 5.7 Structure of the optimization model and dependencies between variables—derived from Rippel et al. (2020b) © 2020 IEEE

(center top). The optimizer noted each round-trip's respective routes, i.e., route three for the first trip and route seven for the second, using the binary encoding.

$$J = -100 \cdot \sum_{j=1}^{N^{rt}} X_{j,1}^{freight} + \sum_{r=1}^{2^{N^c}} \Big((\sum_{j=1}^{N^{rt}} Y_{j,r}^{route}) \cdot D_r^{route} \Big)$$
$$+ \sum_{c=1}^{N^c} \Big((\sum_{j=1}^{N^{rt}} X_{j,c}^{setup}) \cdot D_c^{setup} + (\sum_{j=1}^{N^{rt}} X_{j,c}^{freight})$$
$$\cdot (D_c^{load} + D_c^{unload}) \Big) \tag{5.25}$$

Equation (5.25) shows the cost function used by the proposed optimization model. Similar to the formulation of the scheduling model, this cost function uses a penalty term (first term) to provide the optimizer with an incentive to fetch as many components as possible. Therefore, it multiplies the number of full component sets obtained with a large, arbitrarily chosen negative integer. The remaining terms add the routes' costs and the costs for setup, loading, and unloading operations required for the cycle.

As the model currently uses durations instead of monetary costs, the algorithm obtains the required values for the scheduling model from the results by calculating the duration without the penalty term for RES^{frq} and summing the obtained number of components RES^{amt}. Nevertheless, the algorithm can additionally obtain the sequence of visited ports for each resupply cycle by concatenating the routes obtained by the TSP.

5.4.1.3 Constraints

The stated problem constitutes a version of a multi-periodic Knapsack problem, as described earlier. In consequence, the problem includes several constraints to ensure capacity restrictions but also several constraints relating to the actual scenario and the described structure of the model and its dependent variables.

$$\sum_{j=1}^{N^{rt}} X_{j,c}^{freight} = \sum_{j=1}^{N^{rt}} X_{j,(c-1)}^{freight} \qquad \forall c, \quad 2 \le c \le N^c \tag{5.26}$$

Constraint (5.26) ensures that only full sets are delivered within a cycle. Therefore, it enforces that the number of delivered sets of blades equals the number of towers. Additionally, the number of nacelles equals the number of delivered sets of blades.

$$\sum_{c=1}^{N^c} \left(X_{j,c}^{freight} \cdot REQ_c^{payload} \right) \leq CAP^{payload} \qquad \forall j, \quad j \leq N^{rt} \tag{5.27}$$

$$\sum_{c=1}^{N^c} \left(X_{j,c}^{freight} \cdot REQ_c^{area} \right) \leq CAP^{area} \qquad \forall j, \quad j \leq N^{rt} \tag{5.28}$$

Constraints (5.27) and (5.28) require the payload in all round-trips j not to exceed the vessel's maximum capacity, both in terms of the deck area and the payload.

$$X_{j,c}^{setup} \geq X_{j,c}^{freight} - X_{(j-1),c}^{freight} \qquad \forall j,c \quad 2 \leq j \leq N^{rt}, c \leq N^c \tag{5.29}$$

$$X_{j,c}^{setup} \geq X_{(j-1),c}^{freight} - X_{j,c}^{freight} \qquad \forall j,c \quad 2 \leq j \leq N^{rt}, c \leq N^c \tag{5.30}$$

Constraints (5.29) and (5.30) force the optimizer to note the absolute difference of loaded components between two round-trips. These constraints require the noted value to be larger or equal to $A - B$ and $B - A$, always resulting in a positive valued outcome.

$$X_{j,c}^{freight} \leq Y_{j,c}^{port} \cdot CAP_c^{max} \qquad \forall j,c \quad j \leq N^{rt}, c \leq C \tag{5.31}$$

$$X_{j,c}^{freight} \geq Y_{j,c}^{port} \qquad \forall j,c \quad j \leq N^{rt}, c \leq C \tag{5.32}$$

Constraints (5.31) and (5.32) require the optimizer to denote the type of transported components, i.e., the visited production ports, for each round-trip j as part of the dependent, binary variable $Y_{j,c}^{port}$. Constraint (5.31) limits the maximum number of each type of component to a precalculated value. While redundant to the constraints (5.27) and (5.28), experiments have shown that this constraint acts as relaxation support and can speed up the optimization. The algorithm calculates the maximum number of loadable components as $CAP_c^{max} = min(\frac{CAP^{payload}}{REQ_c^{payload}}, \frac{CAP^{area}}{REQ_c^{area}})$ for all components c.

$$\sum_{c=1}^{N^c}(Y_{j,c}^{port} \cdot 2^{c-1})$$
$$\leq (r-2) \cdot z_{j,r}^{-} + (r-1) \cdot Y_{j,r}^{route} + UB \cdot z_{j,r}^{+} \quad \forall j, r \quad j \leq N^{rt}, r \leq 2^{N^c} \tag{5.33}$$

$$\sum_{c=1}^{N^c}(Y_{j,c}^{port} \cdot 2^{c-1})$$
$$\geq LB \cdot z_{j,r}^{-} + (r-1) \cdot Y_{j,r}^{route} + r \cdot z_{j,r}^{+} \quad \forall j, r \quad j \leq N^{rt}, r \leq 2^{N^c} \tag{5.34}$$

Finally, constraints (5.33) and (5.34) map the combination of visited ports to a unique index for each route. As described earlier and given in the left-hand-side of the constraints, these constraints use a binary encoding of $Y_{j,c}^{port}$ to assign a unique route index $1 \leq r \leq 2^{N^c}$. The constraints use a variant of a so-called Big-M formulation and use two support variables $z_{j,r}^{+}$, and $z_{j,r}^{-}$. These variables denote if the index r is larger or smaller than the round-trip's encoded route-index. In combination, these constraints enforce that $Y_{j,r}^{route}$ only becomes 1, if this calculated index and r are equal. Both constraints use $LB = 0$ and $UB = (2^{N^c}) - 1$ as lower and upper bounds.

5.4.2 Base Port Capacity Planning

The third optimization model presented in this thesis focuses on optimizing the base port capacity on a strategic/tactical level. As noted in the introduction, the installed capacity of offshore wind farms shows a close to exponential increase over the last years. Among others, this trend results from an increasing number of simultaneous offshore projects that will most likely continue increasing as decommissioning and refurbishing projects start emerging (Beinke et al., 2020a). In consequence, the efficient use of port-side resources will become increasingly important.

In practice, the base port's storage spaces need to be planned and chartered comparably long before the actual installation starts. The presented model aims to minimize the chartered spaces (capacity) and the initial inventory levels required to maintain a high installation efficiency. While the actual capacity optimization uses a straightforward formulation, it relies on the results of both of the previously presented models to estimate the in- and outputs for a given project.

5.4.2.1 General Algorithm and Variables

The presented optimization model follows the same assumptions given above for the transport optimization. Consequently, the model's formulation follows a similar structure. Therefore, the capacity optimization assumes that the base port receives a constant resupply of RES^{amt} many component sets with a frequency of RES^{frq}. It expresses this resupply as time series in_l. Thus, each value in this series denotes how many sets of components become available at the base port during time interval l, $l \in \mathbb{N}^+, l \leq N^{rc}$. Similarly, the model uses another time series out_l that describes how many sets of components leave the base port's storage, i.e., will be installed during the respective time interval. Using these two parameters, the optimization model tracks the current inventory levels using the variable $X_l^{invCurrent}$. This variable depends on the noted time series and the optimizer's decision variables $X^{capacity}$ and $X^{initial}$, which denote the selected capacity and initial inventory levels. Moreover, the model uses another parameter INV^- denoting the minimum desired inventory throughout the time series. The algorithm calculates the number of time intervals or resupply cycles N^{rc} within an installation project as $N^{rc} = \frac{N^{owt} - X^{initial}}{RES^{amt}}$ or assumes the worst-case $N^{rc} = \frac{N^{owt}}{RES^{amt}}$.

According to this structure, the model relies on the results of the transport optimization presented earlier to determine the parameters RES^{frq} and RES^{amt}, which it requires to construct the input stream in_l. While it assumes that these values remain constant and do not vary due to weather influences, the output stream depends on estimated weather conditions for the given time intervals. As a result of this planning task's strategic nature, the model cannot rely on current measurements or reliable forecasts to estimate these outputs. Consequently, this offline optimization needs historical data to estimate the number of buildable turbines per time interval. Therefore, this thesis proposes two possible approaches to estimate the output stream. First, a simple approach that analytically aggregates the mean duration of operations. Second, an approach that uses the described scheduling model to simulate an installation project using aggregated historical data instead of using current weather measurements and forecasts. Sections 5.4.2.3 and 5.4.2.4 describe these approaches in more detail. Finally, the results of the capacity optimization act as parameters to the online scheduling for the DSS. Figure 5.8 shows the dependencies between the models and their in- and outputs.

The algorithm first preprocesses the historical dataset of weather measurements. It calculates how long each operation took if it would have been started at a given hour within the dataset using the duration estimation methods presented before. In this case, the algorithm uses the recorded values instead of uncertain forecasts, resulting in a fixed combination of waiting and processing times for each hour in the dataset

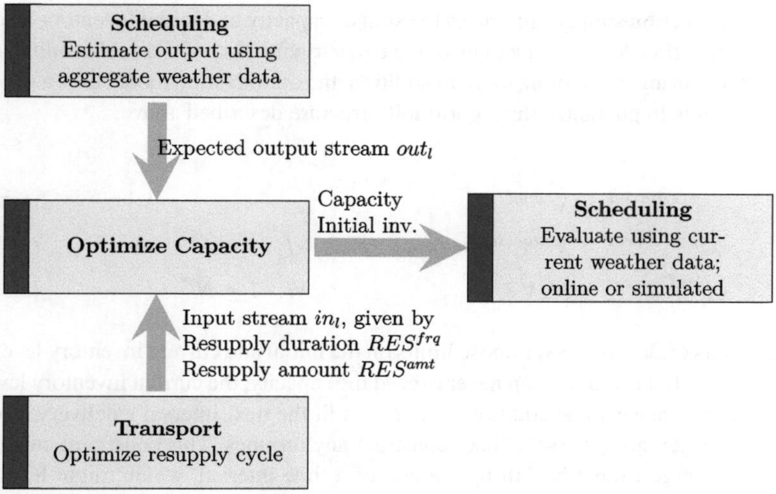

Fig. 5.8 Model dependencies for the capacity optimization

without any uncertainty. While the algorithm could perform these calculations on-the-fly, the static nature of the historical data allows preprocessing the data resulting in a gain of computational speed. Moreover, both estimators use a predefined number of historical years $N^{years}, N^{years} \in \mathbb{N}^+$ for the output estimation. Following the evaluation given later on in Sect. 8.1.2, the capacity optimization generally uses 20 historical years to estimate the output streams.

5.4.2.2 Cost Function and Constraints

The optimization model minimizes the capacity and initial inventory level. It tries to ensure the following conditions: First, a new delivery can fit the storage to prevent delays within the resupply cycle at any interval. Second, the storage contains enough sets to satisfy the installation's demand to prevent delays on the supply chain's installation side at any time. Generally, the model minimizes the cost function given in equation (5.35).

$$J = X^{capacity} + X^{initial} \tag{5.35}$$

This cost function simply minimizes the sum of capacity and initial inventory level. This formulation does not favor any of the two variables, as both variables influence each other during the optimization. In addition, the optimization model uses a series of constraints to guarantee the algorithmic structure described above.

$$X^{initial} \leq X^{capacity} \tag{5.36}$$

$$X_l^{invCurrent} \leq X^{capacity} - in_{l+1} \qquad \forall l, \ l \leq N^{rc} - 1 \tag{5.37}$$

$$X_l^{invCurrent} \geq INV^- \qquad \forall l, \ l \leq N^{rc} \tag{5.38}$$

Constraints (5.36) to (5.38) impose limits on the initial and current inventory levels. While the initial inventory can never exceed the capacity, the current inventory level also needs to leave some additional margin to fit the next interval's delivery, even when the installation vessel cannot construct any turbines. This constraint ensures that the storage cannot be full by the end of a time interval, which might lead to delays in the resupply of components. Finally, the last constraint requires the current inventory to never drop below a specified threshold. Choosing $INV^- = 0$ ensures that the storage cannot run empty during the installation. Higher values induce a specified safety stock to counteract unpredicted peaks in the installation's demand due to good weather periods.

$$X_1^{invCurrent} = X^{initial} - out_1 \tag{5.39}$$

$$X_l^{invCurrent} = X_{l-1}^{invCurrent} + in_l - out_l \qquad \forall l, \ 2 \leq l \leq N^{rc} \tag{5.40}$$

The final two constraints (5.39) and (5.40) describe the current inventory's progression over the time intervals. Therefore, the first constraint ensures that the first interval's storage equals the initial inventory minus all turbines installed during this period. The inventory for all future intervals depends on the last interval, plus the number of component sets delivered, minus the number of turbines installed. This model shows a clear dependency on the respective in- and output rates for each interval. Therefore, the next sections propose two methods to estimate the output rates using historical data.

5.4.2.3 Output Estimation by Aggregation

The first estimator proposed in this thesis uses the dataset directly to construct the average duration of each operation over the required period. Afterward, it calculates the number of installable turbines for each interval l to obtain the output stream out_l.

In a first step, it selects the time window from the dataset, starting at the respective date and time for each interval l. The selected window has a size conforming to the resupply cycle's duration RES^{frq} and the selected number of past years N^{years}. Figure 5.9 schematically shows the time window used to construct the mean duration for an installation operation. The figure shows that both parameters, the resupply frequency and the number of historical years, determine the level of aggregation applied by this estimator. Thereby, shorter windows would result in a more detailed representation of the (mean) weather conditions, while larger windows imply a stronger smoothing.

Hour=	0	1	2	3	4	5	6	...	336
1999	21	20	19	19	19	19	19	...	20
...	...	...	...	...	...	...	...	...	...
1979	26	25	24	23	22	21	20	...	19

Fig. 5.9 Example for the time window used to build mean values in the aggregation-based estimator for $RES^{frq} = 336$ and $N^{years} = 20$

The second step calculates the mean duration for each relevant offshore operation, denoted as d_l^{owt}, d_l^{load}, and d_l^{move} for this particular time window. Finally, it calculates the number of installable turbines for this interval, as given in equation (5.41).

$$out_l = \frac{RES^{frq}}{d_l^{owt} + d_l^{load} + (d_l^{move} \cdot \frac{2}{CAP_v})} \tag{5.41}$$

Generally, this equation calculates the process chain's mean duration by dividing the length of the respective interval by the sum of all operations required to perform an installation. The equation calculates the movement time as two trips, there and back, divided by the number of turbines the vessel could transport based on its capacity.

5.4.2.4 Output Estimation by Scheduling

The second estimator proposed in this thesis aims to provide a higher level of detail while estimating the number of installable turbines per interval. For each interval l, this estimator uses the scheduling model to obtain an optimized schedule. The planning horizon equals RES^{frq}, whereas the optimizer assumes that all vessels have already fully loaded and moved to the installation site. These assumptions reduce the chance to underestimate the number of installable turbines, which might result in underestimating the initial inventory level. The scheduling model relies on the durations for each operation $D_{o,k}$. Therefore, this estimator constructs the mean duration for each operation o and each hour k within the current interval l separately.

Hour=	0	1	2	3	4	5	6	...	336
1999	21	20	19	19	19	19	19	...	20
...	...	...	...	...	...	...	...	...	...
1979	26	25	24	23	22	21	20	...	19

Fig. 5.10 Example for the time window used to build mean values in the scheduling-based estimator for $RES^{frq} = 336$ and $N^{years} = 20$

Figure 5.10 schematically shows the time window used to construct the mean duration for an installation operation. This procedure increases the level of detail by focusing on single hours instead of the whole planning horizon. After computing the optimal schedule for each interval l, the estimator constructs the output stream out_l by counting the number of installation operations within each schedule.

5.5 User Interface

The methodology requires several parameters that either describe the scenario or affect the inner workings of the methodology. The methodology specifies these parameters and their required format, e.g., as part of the state variable. When applied as online DSS, the real-world systems provide most of these parameters using dedicated interfaces, which this thesis considers out-of-scope. Nevertheless, to conduct simulation experiments, e.g., for the capacity optimization or during the evaluations later on in chapter 8, the methodology requires a manual specification of these parameters in the correct format. As can be seen from the state description in Table 5.1, even the scenario's parameters might pose a challenge to process experts to specify in the correct format. In contrast, one of the functional requirements (R4) states that the methodology should provide an intuitive way for process experts and project planners to apply it.

Therefore, this thesis proposes to decouple the inputs required by the methodology and its mathematical formulation from the scenario specification. On the one hand, this decoupling allows process and methods experts to use more intuitive interfaces to specify the scenario and modify the methodology's configuration. On the other hand, this decoupling ensures the correctness of the generated inputs without process experts knowing about these formats. Consequently, this thesis proposes meta-modeling techniques and model transformations to generate the required

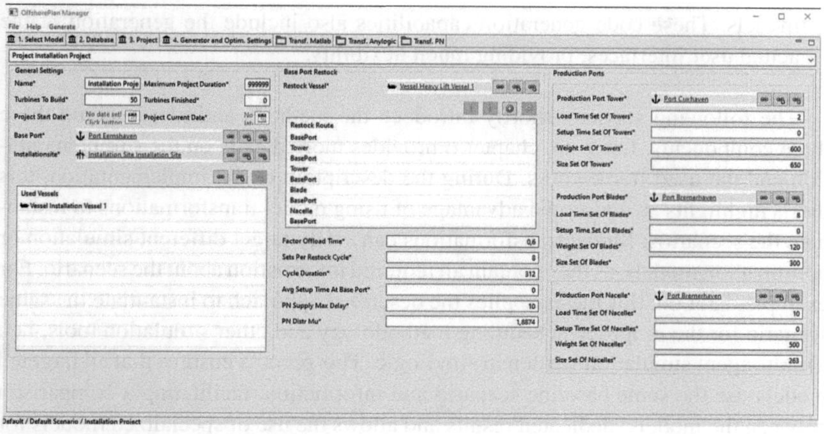

Fig. 5.11 Screenshot of the tool's generated user interface

inputs from a more general, intuitive scenario model. While this thesis does not focus on the design of user interfaces, it exemplary shows an initial, generated interface that allows the scenario specification and includes method-specific parameters, as shown in Fig. 5.11. In summary, the decoupling of the methodology and the user interfaces includes three distinct elements:

Data model / meta-model: This approach provides users with an intuitive model to define the scenario and its parameters in the form of user-specific data models. Each of these models implements a custom meta-model, which describes the structure of the user models. Section 5.5.1 describes the meta-model and its deduction. This thesis applies the OMG's EMOF standard to define the meta-model.

Model transformation: This approach extracts the required information from the user-specific data models and applies model-to-text transformations to generate viable inputs for the methodology. This thesis applies the OMG's M2T standard to generate scenario files that the methodology can directly load and execute. The model transformations ensure that the required information will be converted correctly and maintain the correct format.

User Interface: This thesis relies on implementations of the noted standards that are part of the EMF. In addition to providing these implementations, EMF also provides means to generate software tools, able to create and maintain user models. These code generation capabilities also include the generation of the actual user interfaces, providing much flexibility.

The following sections shortly introduce the methods and ideas behind these three components. Later on, chapter 6 provides more details on their implementation and the used frameworks. During the description of the implementation, this thesis highlights an additional advantage of using model transformations to instantiate the scenarios: Model transformations can easily target different simulation or optimization models, as they contain all required information about the scenario. For example, later on, this thesis applies the described approach to instantiate the same scenario for the proposed scheduling methodology and other simulation tools, i.e., a multi-agent simulation written in AnyLogic. This process ensures that all targeted models use the same baseline scenario and information, facilitating a comparison between the models' dedicated results and allows the use of specialized models for different planning tasks.

5.5.1 Meta-Model

The meta-model serves to describe the structure and contents of the later user models. Therefore, it needs to include all relevant information to plan and simulate an offshore installation process. Furthermore, the meta-model needs to structure this information in an accessible way to facilitate its use by process experts. This section shortly summarizes the meta-model and its deduction from the literature. A detailed description of the literature analysis and the resulting model has been published in Rippel et al. (2019c).

In general, the meta-model constitutes a domain model for the installation of OWFs, describing all relevant parameters and their interconnections. Literature states that no standardized procedure for the development of such domain models exists, but Stuckenschmidt (2011) presents some best practices for developing a generalized domain model. Therefore, the author proposes a series of six steps, as shown in Fig. 5.12.

The first step aims to define the focus of the domain model under development by clearly selecting which aspects to include and neglect. In this thesis, the installation of OWFs constitutes the area of application. Though Sect. 5.1 already stated its limitations, the meta-model can include additional parameters and concepts, even though the scheduling methodology will disregard them at this point.

The second step focuses on the reuse and extension of existing models. While there exists no generalized domain model for the installation of OWFs, the state of

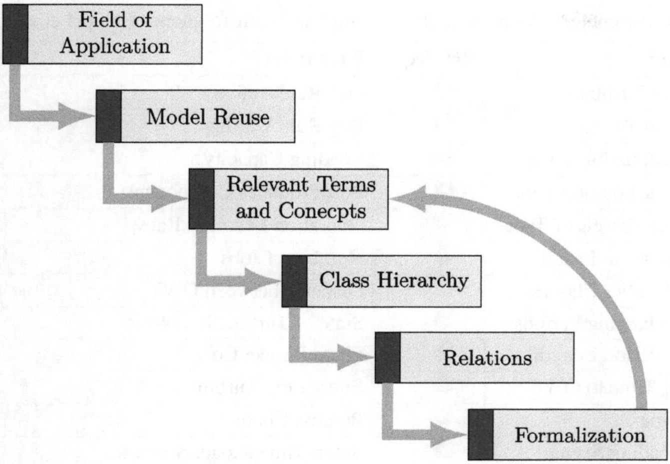

Fig. 5.12 Procedure to model domain ontologies, according to Stuckenschmidt (2011)—adapted from Rippel et al. (2019c)

the art in Sect. 4.2 presents several models and tools that focus on the installation process and serves as a baseline for this step. An analysis of these models resulted in a list of all used parameters and variables.

The third step aims to identify common terms and concepts within the domain. This step analyzed the list of parameters and concepts obtained before by aggregating parameters with the same meaning. Authors often describe the same concepts within different models, but use different notations, names, or even modeling concepts. By aggregating these different notations, the third step provided the parameters listed in Table 5.3. The table lists each (aggregate) parameter identified during the literature analysis and shows its frequency of occurrence. The frequency distinguishes between parameters that occur in more than 70% of the analyzed models, parameters occurring in 30% to 70%, and parameters, which authors use in less than 30% of the model.

The last three steps aim to formalize the identified terms and concepts into a formally defined structure. As noted before, the meta-model developed as part of this thesis uses the OMG's EMOF as its meta-meta-model. The EMOF's notation closely resembles the notation of class diagrams used in the UML. Basically, it consists of classes, attributes, and methods. In fact, the EMOF serves as meta-model for the UML and, thus, also for class diagrams. The final steps, i.e., the construction

Table 5.3 Parameters identified in the literature and their frequency (Rippel et al., 2019c)

Parameter	Rel.Frq.	Parameter	Rel.Frq.
Number of Turbines	●	Day Rate Active	◔
Component Type	●	Day Rate Waiting	◔
Comp. Installation Time	●	Loading Capacity	◔
Component Loading Time	●	Port Produces Component	◔
Number of Vessels of Type	◕	Operation Learning Rate	◔
Req. Weather to Install	◕	Number of Jobs	◔
Seq. of Weather Classes	◕	Distance between OWT	◔
Num. of Planning Periods	◕	State of Turbine in OWF	◔
Planning Period Length	◕	Fixed Project Cost	◔
Traveling Time to OWF	◕	Energy Per Turbine	◔
Vessel Type	◕	Process Chain	◔
Vessel Loading Scenarios	◕	Setup Time (Load. Scenario)	◔
Required Weather to Load	◕	Setup Cost (Load. Scenario)	◔
Timeseries of Weather Data	◕	Seafastening Time	◔
Project Start Date	◕	Transshipment Time	◔
Distance to OWF	◕	Jack-up Rate (Time)	◔
Travel Speed	◕	Minimum Renting Period	◔
Port Storage Capacity	◕	Commissioning Cost	◔
Req. Weather Seafasten	◔	Decommissioning Cost	◔
Req. Weather Transship	◔	Port Process Times (Load)	◔
Cost for Vessel in Period	◔	Port Weather Rest. (Load)	◔

● Often, (≥ 70%) ◕ Frequent (≥ 30%) ◔ Rarely (< 30%)

of a class hierarchy, the definition of relations between these classes, and the final formalization in EMOF resulted in the class diagram given in Figs. 5.13 and 5.14. In addition to the parameters given in the literature, the diagram in this figure also includes all relevant concepts and parameters required for the models proposed in this thesis.

In its current state, the proposed meta-model omits the inclusion of weather data and fixed project or vessel costs, e.g., in terms of commissioning or decommissioning costs. The proposed scheduling methodology focuses on variable costs and does not consider fixed costs, as they do not depend on the planning. Moreover, sensible integration of these costs requires a suitable cost model, which this thesis considers

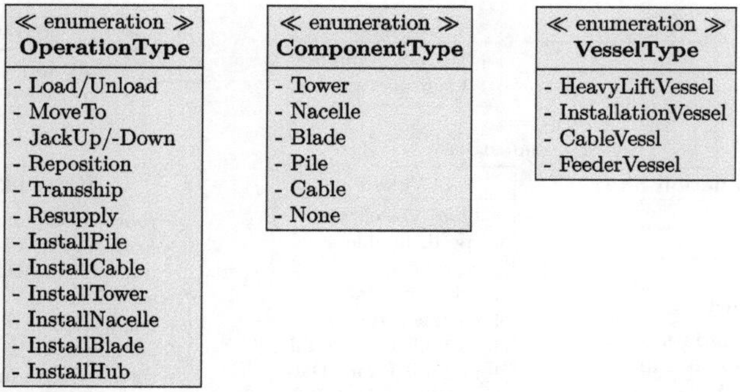

Fig. 5.13 Enumerations used in the meta-model

out-of-scope. In the context of weather data, simulation and optimization models in the literature usually rely on specialized representations of weather conditions. For example, the proposed methodology uses hourly measurements. Other approaches use probability distributions or sequences of weather classes. While conversion of weather data might be possible to a certain degree, this thesis assumes that each optimization or simulation model provides and manages its own weather data.

Figure 5.13 shows the types of components, vessels, and operations included in the meta-model. As noted above, this model also allows the definition of processes that this thesis considers out-of-scope, e.g., installing founding structures or the use of feeder vessels. To a certain degree, the proposed models can handle such processes. For example, the installation of founding structures only requires defining the process chains and operations. In contrast, the use of feeder strategies would require reworking at least the optimization model's constraints.

The class diagram in Fig. 5.14 shows the derived meta-model with a focus on its contents. Therefore, the diagram omits containment-associations (UML-Aggregation or UML-Composition), which indicate where the model actually stores information. In this context, the scenario acts as a container that, in terms of a database, stores all other model elements, e.g., instances of vessels, ports, operations, or different instances of resupply cycles. Each general association shown in the diagram depicts a reference to one of these stored model elements, which allows users to exchange predefined elements quickly and, e.g., experiment with other vessels or ports. In addition, the figure only shows a simplified version of the class algorithm settings to improve its readability. The meta-model covers the following classes or aspects of the installation process:

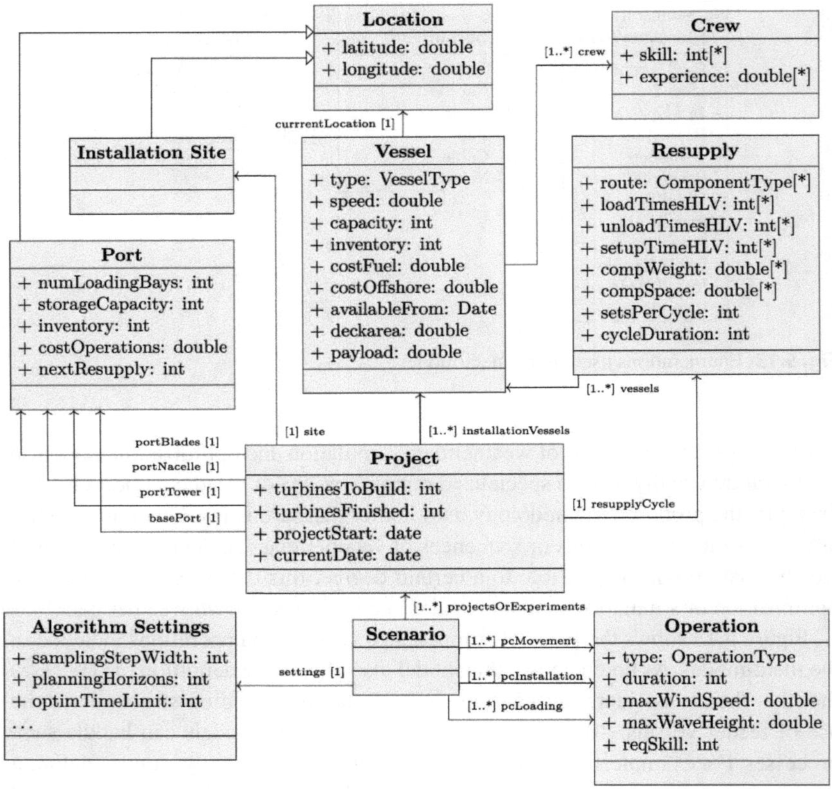

Fig. 5.14 Meta-model derived from literature and the proposed models—Extended and restructured from Rippel et al. (2019c)

Scenario: The scenario acts as the main entry class of the model and constitutes its primary database. The scenario allows users to define arbitrary numbers of other model elements, e.g., different vessels they might want to use in an installation project or different projects that use resources from the same pool.

Algorithm Settings: This class defines all relevant parameters used to control the overall scheduling methodology or the methods it includes. These include MPC related settings like the desired sampling step width, optimizer settings, like the maximum optimization time or the allowed gap, or the robustness parameter ω used by the sliding window duration estimation. The meta-model assigns these settings to the scenario class to ensure that all simulation experiments use the same settings unless globally changed.

Operation: The operation class describes basic operations, e.g., given in Table 2.2. It uses the previously introduced enumeration to define the type of operation and subsumes its duration and weather limits. The scenario uses operations to define the sequence of basic operations for each aggregate operation, as described before.

Project: Instances of the class project describe distinctive installation projects. The class itself manages the number of finished turbines, turbines to build, and the projects starting and current date. By referencing elements of other classes, it defines, e.g., which vessels the planner aims to use in this project, or which of the available ports serve as production or base ports.

Resupply: This class comprises all parameters to describe the projects' planned resupply cycle. On the one hand, this class includes detailed parameters for planning these cycles, like processing times or the space and weight requirements of components. On the other hand, it also manages the resulting cycles, e.g., by storing the sequence of visited ports or the cycle's duration.

Location / Installation Site: The location class acts as the base class for all locations, i.e., ports and installation sites and describes their latitude and longitude.

Port: In contrast to the installation site, this class contains several parameters to characterize base or production ports. These include details on the port's storage or the number of loading bays.

Vessel: The class vessel describes the most important characteristics of vessels in terms of transport optimization and scheduling. These characteristics include hourly costs, capacities, and traveling speeds used for the calculation of movement operations. The class uses the previously introduced enumeration `VesselType` to differentiate between several types of vessels. Nevertheless, all of these types comprise the same attributes.

Crew: Finally, this class describes the abilities and experience of the personnel. While the current methodology does not consider these model elements, models in literature use these parameters to modify the duration of operations (learning rate) or assign personnel to operations (workforce scheduling).

Using this model, process experts can define their installation scenario and the involved resources in a comparably intuitive way. Moreover, this meta-model allows the development of specialized user interfaces, e.g., by hiding or separating method-specific settings from the actual scenario. This meta-model contains most of the information derived from literature and all relevant information to enable the generation of simulation experiments for the proposed methodology, e.g., as used later on during the evaluations presented in chapter 8.2.

5.5.2 Model Transformation

The meta-model allows users to define their scenarios and parameterize the methodology using distinct user models, allowing them to create and manage several different scenarios, projects, or configurations. The methodology applies model transformations to create the necessary inputs for the methodology. Therefore, the transformations create a state description that the methodology can load and execute, e.g., as an initial state for an online optimization or optimize transport routes. This thesis proposes to generate source code that allows a generation of the state in the computer's memory. The OMG's M2T standard provides a suitable tool to convert EMOF-models into text, e.g., into source code.

The M2T standard describes the use of templates. For each class of the meta-model, the standard allows defining a template describing how instances, i.e., objects, of each class should be represented as text. The methodology saves the resulting texts as an executable code file that the optimization models can load on initialization. Figure 5.15 shows an example of such a template on top and the generated code snippet on the bottom. This example references two objects: an instance of the class Project as `proj` that the template targets, and its parent or container class, the scenario as `scen`. The template generates plain text directly, e.g., the comments or variable names. In addition, the M2T standard proposes the use of the OMG's OCL to enable complex calculations on the referenced objects. These templates envelop OCL-statements with square brackets, i.e., in between `[` and `/]` as known from other markup languages. For example, the example's first statement `[proj.maximumProjectDuration/]` denotes that the template should print the value of the variable `maximumProjectDuration` of the object `proj` at this location in the generated text. Moreover, the example shows a more advanced statement. This for-statement first obtains all operation objects from a list (`scen.Operations`) and prints their names. It separates the names with commas. Finally, it precedes the generated text with the variables name and Matlab's list qualifier (`'scenario.opName = ['`) and similarly closes the list by appending closing brackets and a semicolon. OCL allows to include complex calculations, function calls, or list operations to enable arbitrary templates.

These templates ensure that the transformation always applies the same rules and procedures when converting the user model into an executable source code. Moreover, they ensure that the format of the resulting file always conforms to the

```
%%%% Scenario - Limits %%%%
scenario.targetPlanTime = [proj.maximumProjectDuration/];
scenario.OWTsToBuild = [proj.turbinesToBuild/];

%%%% Scenario - Process data %%%%
[for (op : BaseOperation | scen.Operations)
   before ('scenario.opName = [') separator (',\t') after ('];')]
      "[op.name/]"
[/for]
```

```
%%%% Scenario - Limits %%%%
scenario.targetPlanTime = -1;
scenario.OWTsToBuild = 50;

%%%% Scenario - Process data %%%%
scenario.opName = ["Install Tower", "Install Nacelle", "Install
   Blade 1",
   "Install Blade 2", "Install Blade 3", "Install Hub", "Move",
   "Reposition in Field", "Jack-up", "Jack-Down",
   "Load OWT Components", "Install OWT"];
```

Fig. 5.15 Excerpt from an Acceleo-M2T template and generated Matlab code

specifications. The use of such templates allows users to target different executable models by adding additional templates. This process allows users to generate executable models for the same scenario, e.g., for the presented scheduling methodology, but also other simulation or optimization tools, as long as they support reading their initial state from a structured text file.

5.5.3 Generation of User Interface

Similar to the generation of executable models, the meta-model enables the (semi-automatic) creation of specialized user interfaces, allowing for different views on the user model. While this thesis considers the actual interface design out-of-scope, it briefly presents an example of such a user interface in the next chapter in Sect. 6.3.3. The user interface provides a dedicated form for each modeling element (class) specified in the meta-model. The design process can utilize another model transformation, in this case, Eclipse Forms, to specify and generate source code for each form based on the underlying meta-model.

Prototypical Implementation 6

This chapter shortly presents the main aspects of the DSS's implementation used for this thesis. Therefore, it first presents relevant aspects of the implementation of the underlying methodology itself in Sect. 6.1. These aspects mainly cover different settings and modes for the optimizer. Afterward, this section introduces a multi-agent simulation (MAS) model that serves as a benchmark for other decision strategies in Sect. 6.2. Such strategies include, e.g., a simple priority-based strategy or a strategy that mimics the current ad-hoc decision making performed in practice. Finally, this section provides additional implementation details for the model transformation framework in Sect. 6.3. This thesis uses this framework to set-up and manage different scenarios to ensure equality for scenarios in the proposed MPC-based approach and the MAS.

6.1 Scheduling and Optimization Methodology

This thesis uses an implementation of the DSS and all of its components in MATLAB 2018b. Furthermore, it uses CPLEX 12.10 to solve the optimizations. Therefore, it applies Matlab's problem-based formulation for optimization problems, facilitating a straightforward implementation of the constraints and rendering the code more readable. Compared to the default solver-based formulation applied by most tools, the problem-based specification allows for a very similar structure and notation, as presented in the last section. The listing in Fig. 6.1 shows a code example for the constraints (5.6) to (5.8). The figure shows that each *for-all* clause (∀) directly translates into a `for-each` loop, while the actual equations translate directly into the respective Matlab code.

© The Author(s), under exclusive license to Springer Fachmedien Wiesbaden GmbH, part of Springer Nature 2026

D. Rippel, *Online Decision Support for Offshore Wind Farm Installations*,
https://doi.org/10.1007/978-3-658-49912-9_6

```
% Optimization Variables
location = optimvar('location', vI, N, 'Type', 'integer',...
         'LowerBound', 0, 'UpperBound', 1);
xOWT = optimvar('xOWT', vI, N, 'Type', 'integer',...
         'LowerBound', 0, 'UpperBound', 1);
[...]
% Constraints: Location
for v = 1:vI
    for k = 1:N
        if( k > 1)
            constr_location1(v,k) = location(v,k) == ...
                location(v,k-1) - (xToPort(v,k)) + xToSite(v, k);
        else
            constr_location1(v,k) = location(v,k) == ...
                scenario.currentLocation(k,v) - (xToPort(v,k)) ...
                + xToSite(v, k);
        end
        constr_location2(v,k) = location(v,k) >= xOWT(v,k);
        constr_location3(v,k) = location(v,k) <= (1-xPortOp(v,k));
    end
end
```

Fig. 6.1 Constraints (5.6)–(5.8) in Matlab's problem-based formulation

After specifying the optimization problem, Matlab forwards the generated problem-structure to CPLEX 12.10 using CPLEX's built-in Matlab API. After solving the problem using CPLEX, Matlab obtains the result vector and transforms it back into its internal structure, which it further processes as described in the last chapter.

6.1.1 CPLEX Configuration

Internally, CPLEX uses a branch-and-bound algorithm to obtain exact solutions for the optimization problem. This algorithm spans a search tree over the decision variables. This algorithm obtains a relaxed solution to the overall problem as an initial point in the first step. This relaxation provides a solution to the problem without regarding the imposed integer constraints. Such a relaxed solution always imposes a lower bound, as no integer-constrained solution can result in lower costs than the unconstrained solution. Afterward, the algorithm iteratively selects variables and creates two candidates for each by branching the current value upwards or downwards and reimposing the related integer-constraints. At this, it reevaluates the current lower-bound by reintroducing additional integer constraints on the decision variables. Finally, it selects candidates for evaluation. At this stage, it verifies if these candidates satisfy all constraints and calculates the cost. If such a candidate represents an integer-based solution and has lower costs than the current incumbent (upper bound), it selects this candidate as new incumbent. During this process, the algorithm prunes all sub-trees that cannot result in lower costs as the current incumbent, e.g., because their relaxations already have higher or equal costs. Moreover, it prunes sub-trees that have already become infeasible. The algorithm repeats this process until the gap between the current incumbent and the lower bound falls within a given threshold.

CPLEX offers a multitude of parameters to modify the behavior of this algorithm. This section shortly summarizes the most notable changes to the default parameters used in this thesis. Most of these settings resulted from empirical tests that aimed to speed up the methodology's solution speed.

Aggressive node presolve: This parameter determines how much time CPLEX spends during the problem relaxation and presolving stage. As this stage generally takes less than a second, aggressive presolving allows CPLEX to obtain better solutions early on. Moreover, this thesis uses settings for strong and deep probing during the optimization. These settings instruct CPLEX to invest additional time to scan for possible incumbents close to the initial solution.

Search mode: Due to the combinatorial nature of the problem and its non-monotone cost function, this thesis instructs CPLEX to search for as many "hidden feasible" solutions as possible. This instruction results in a broad-first search within the solution space. In contrast to a depth-first search, if set to "optimality", CPLEX obtains viable solutions faster. Experiments showed a faster convergence towards the optimum, as CPLEX can prune larger parts of the search space with every new incumbent. Moreover, this thesis instructs CPLEX to select candidate nodes using its internal "expect decreased costs" heuristic, compared to, e.g., favoring to decrease the gap between the current incumbent and the lower bound for the same reason. With similar reasoning, the implementation instructs CPLEX to favor candidates that branch upwards, often resulting in pruning large sub-trees early on due to infeasible solutions. Finally, the implementation instructs CPLEX to use its "opportunistic search mode" in contrast to the "deterministic" one. This mode allows CPLEX to select non-preferred but promising candidates within the search space randomly. Even though this mode can result in different solutions for each execution, experiments showed a high increase in the solution speed.

Symmetry detection and numerical stability: Finally, the implementation instructs CPLEX to use "aggressive symmetry detection" and numerical problem scaling. The non-monotone nature of the cost function results in symmetrical situations, even though the imposed penalty terms significantly reduce these scenarios. Nevertheless, numerical instable or symmetrical states highly decrease the optimization speed. Consequently, the implementation modifies several settings to reduce the risk of such situations or support their detection. While these settings slightly reduce the overall speed, they prevent deadlocks and generally result in a stable optimization process.

Most of these parameter settings result from the non-monotonous nature of the solution space and the cost function. As a result, the optimizer cannot merely follow a distinct gradient as the search space contains several local minima. Consequently, the presented parameters prefer a broader search strategy to omit such local minima.

6.1.2 Optimization Modes

This thesis uses two different optimization modes during its evaluation. While the default optimization mode limits the time that CPLEX can use for the optimization and returns the final result, the second optimization repeats the optimization until it cannot obtain better solutions. This thesis refers to this second mode using the term stagnation.

Generally, the optimizer obtains very good or optimal solutions with comparably short computational times, as shown later in Sect. 8.2.2. Nevertheless, this NP-hard/NP-complete problem's combinatorial nature prevents CPLEX from proving the optimality of solutions for larger problem instances, e.g., if applying multiple installation vessels. Consequently, the result may or may not be optimal in such cases.

Default: The first optimization mode constitutes the default. Thereby, the implementation allows CPLEX a given amount of time, usually 15 minutes, to find an optimal solution. If the time expires before determining the optimal solution, CPLEX returns the best incumbent.

Stagnation: The second optimization mode follows the same process but uses the last incumbent to repeat the optimization. This way, the optimizer often obtains additional solutions, as the branch-and-bound algorithm heavily depends on the quality of the initial solution. Using the stagnation mode, Matlab first performs a short optimization run of usually 15 minutes to obtain a viable initial solution. Afterward, it repeats the optimization using a time limit of one or two hours and the last best incumbent as initial solution, until repetitions yield no further improvements. While this process still does not guarantee an optimal solution, it highly increases the chances of obtaining a near-optimal or optimal solution, even if CPLEX cannot prove optimality.

6.2 Multi-Agent Simulation Model

This thesis includes a separate implementation of the installation process as a MAS to evaluate the model transformation and compare the behavior of the DSS with other decision strategies. The model has been implemented in AnyLogic 8.5.2 and closely follows the models used and presented in Ait Alla et al. (2017) and Oelker et al. (2018). Figure 6.2 shows a screenshot of the simulation model.

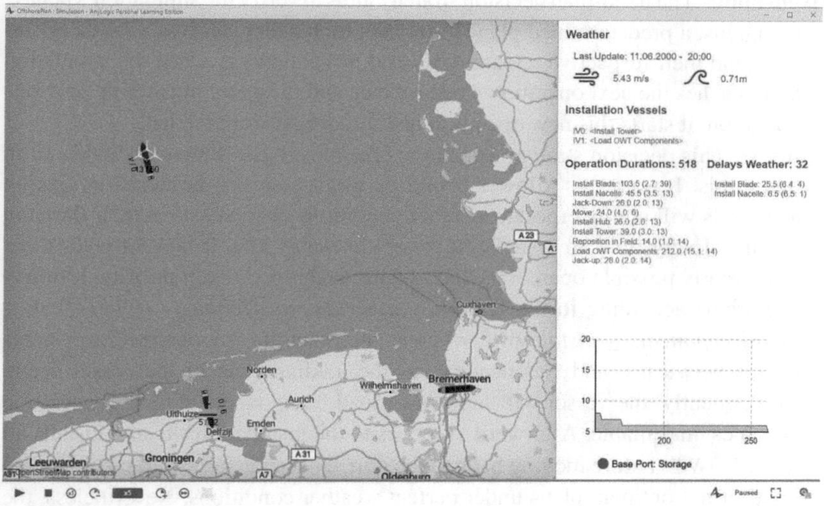

Fig. 6.2 Screenshot of the AnyLogic MAS used as benchmark

Instead of focusing on feeder concepts, this implementation only supports the traditional installation concept. The MAS follows the same mark-up as given in the process description above, i.e., it simulates the installation process with the same restrictions and assumptions as described earlier in this thesis. It consists of agents for installation vessels, HLV, ports, and an agent for the installation site. Moreover, it uses a similar state representation to instantiate its scenario. Thus, at the beginning of the simulation run, it instantiates an agent for each port, using the state's geolocation and capacities using AnyLogic's GIS components. Afterward, it uses the state's information to instantiate installation and HLVs at the provided locations with the given characteristics. This simulation model uses the same notion of aggregate and basic operations as described in the method's concept, i.e., installation vessels can perform either of the aggregate operations *loadComponent*, *moveToSite*, *moveToPort*, or *installOWT*. Additionally, they can *wait* for one hour. Each of these aggregate operations consists of a sequence of operations as described above, each with their respective weather restrictions. If current weather conditions exceed these restrictions, the vessel agent resets its current basic operation and instead waits for one hour. During the installation, each vessel agent selects the operation to perform using one of three implemented methods:

Fixed-Plan: The fixed-plan decision strategy aims to verify the simulation's behavior against a predetermined plan. Therefore, each agent receives a list of operations and their respective release dates and just follows it. Once the simulation time reaches the next operation's release date and the agent finishes its previous operation, it starts this new operation and removes it from the list.

Priority: This decision strategy constitutes the most basic strategy considered in this thesis. It serves as a lower bound or worst-case for the decision making as vessels will not incorporate weather conditions or forecasts in their decision making. Every time an installation vessel becomes idle, it first retrieves a list of currently possible operations and assigns each operation a priority. It orders operations according to the following scheme: $moveToSite = moveToPort \leq loadComponents \leq installOWT$, whereas the exact priority of loadComponents ranges between 0 and 1 depending on the loading state of the vessel agent. Consequently, the vessel will first try to load components until this operation becomes unavailable. Afterward, it moves to the installation site and installs all loaded OWTs before moving back to the port. This strategy would result in very good, if not optimal, plans under perfect weather conditions. Nevertheless, the disregard of weather conditions results in increasingly worse plans if conditions become less optimal.

Forecast: This decision strategy aims to provide a benchmark for the ad-hoc decision making found in practice. Every time an installation vessel reaches the base port, it evaluates the weather forecast for the upcoming two weeks to determine a shortterm plan for the next installation cycle.

Therefore, it first estimates each operation's expected duration within the forecast horizon using the Markov-Estimator presented in Sect. 5.3.3.2. Afterward, it determines possible loading sequences. Therefore, it derives a schedule with the required loading operations to achieve a loading state of n, with $1 \leq n \leq CAPv$, considering its initial current loading state, its maximum capacity CAPv and the other vessel's planned loading operations. Each of these sequences results in the earliest date to start the installation and the number of loaded sets n. The vessel then constructs plans for the actual installation cycle, again planning to install n

turbines using the shortest possible loading sequence for each cycle. The vessel applies the same "operation chaining" described in Sect. 5.3.3 to determine the duration of the cycle beginning at the end of the loading sequence. Each plan introduces a waiting time between the loading and the installation by shifting the start forward one hour as long as the total duration decreases or remains the same. The latter shifts, move expensive offshore delays into cheaper port side waiting times. Finally, it ranks each of these plans by first selecting the plan with the shortest offshore delay, then the plan with the most installations, and finally, the plan with the shortest waiting time.

While a vessel agent has a valid plan, it follows this plan until it reaches the port again, i.e., the planned cycle ends. Nevertheless, if it cannot schedule a loading sequence, e.g. because the base port has no sets stored, the vessel aborts the planning and waits for a given amount of time before retrying (default 24 hours). The same accounts if the first loading operation cannot be scheduled within 72 hours to ensure that suitable plans still fit the two-week prediction horizon. Moreover, if the delay between a planned loading operation and its actual start becomes too large (default > 10 hours), the vessel discards the plan and calculates a new one.

HLV always follow a predetermined resupply cycle. This cycle consists of a series of ports to visit and their related loading and offloading times, as described in the transport optimization problem. Once the HLV reaches the base port with the last delivery of the cycle, it checks the port's storage capacity and either unloads the last components, rendering all sets available to the base port, or waits until the port provides sufficient capacity. Afterward, it starts a new resupply cycle until the port obtained all component sets. The base port agent manages the component inventory and the availability of loading bays.

6.3 User Interface

As noted before, this DSS uses a dedicated tool to specify simulation scenarios, as given in Sect. 5.5.1. This scenario manager tool consists of the implemented meta-model, a user interface, and the respective model transformations. The scenario manager developed as part of this thesis thereby supports the instantiation of experiments for the MPC-based approach and the described AnyLogic MAS described in the last section by applying model transformations. These transformations guarantee that both models use the same scenario as input and that the experiments' results remain comparable.

The implementation of this tool uses the Eclipse Platform to implement the different components. Generally, Eclipse provides a plugin-based platform, called Eclipse Rich Client Platform (RCP), which already provides several plugins to develop modeling languages and model transformations within the EMF. Moreover, the code resulting from these plugins again constitutes another plugin, which the RCP can host and provide to users. Consequently, this framework of plugins allows creating a dedicated software application that provides the generated user interfaces, transformations, and plugins for model management directly to a user. Figure 6.3 shows an overview of the tool's structure and the used plugins and standards for each component.

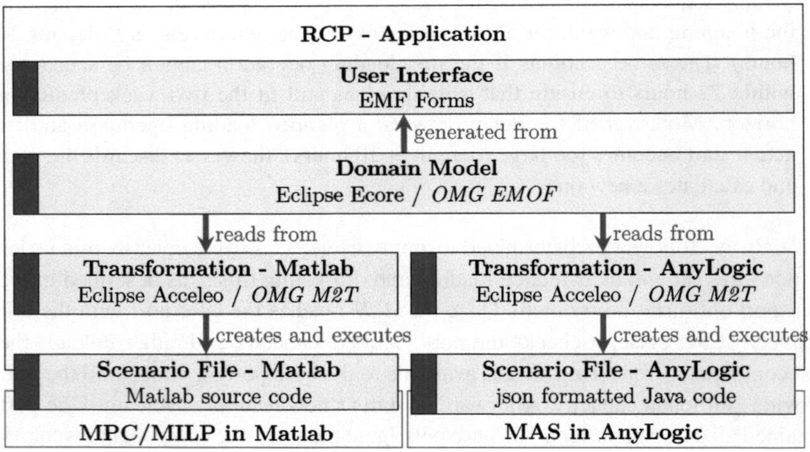

Fig. 6.3 Structure of the scenario tool with applied plugins and standards

6.3.1 Meta Model—Eclipse Ecore

As shown in Fig. 6.3, the tool uses the Eclipse Ecore plugin to define the meta-model and generate the respective plugins to manage, manipulate, load, save, and display the resulting user models. Ecore constitutes an implementation of the EMOF standard and allows defining the meta-model in a class diagram as depicted in the last chapter's Figs. 5.13 and 5.14. In addition to the classes, attributes, and associations given in those figures, the meta-model also includes several containment-associations between the scenario class and most of the other classes. These containments result in the scenario storing all instances of these objects and acting as a database. Consequently, users can define, e.g., several vessels and store them as part of the scenario. Each project then selects vessels from this database and assigns them for usage. This structure facilitates quick experimentation with various resources or network layouts while similarly facilitating the reuse of model elements for different experiments or projects.

6.3.2 Model Transformation—Eclipse Acceleo

The tool provides two different model transformations; both implemented using the Eclipse Acceleo plugin as an implementation of the OMG's M2T standard. Both transformations consist of a single template and follow the same structure, as shown previously in Sect. 5.5.2. Each template targets either the proposed MPC-based approach or the described AnyLogic MAS. In terms of the MPC-based methodology, the template generates Matlab code and saves it as an m-file within Matlab's workspace. Once the algorithm runs, it first executes this code to generate the initial state model if such a file is present. Otherwise, the methodology initializes with a default state, as described in the next chapter. The MAS follows the same procedure. Instead of providing a code file, the transformation provides the simulation model with a serialized Java class in a textual json format. Before instantiating the agents within the simulation, the MAS checks if it can locate such a file within its workspace and either overrides its default scenario by deserializing it or proceeds with the default scenario.

6.3.3 User Interface—EMF Forms

Finally, this implementation uses the Eclipse—EMF Forms plugin to generate a user interface based on the defined meta-model. This plugin allows to define and layout interface components for each of the meta-model's classes, as indicated in Fig. 6.4. This figure shows the definition for the project class on the left-hand side and an exert of the generated user interface on the right-hand side.

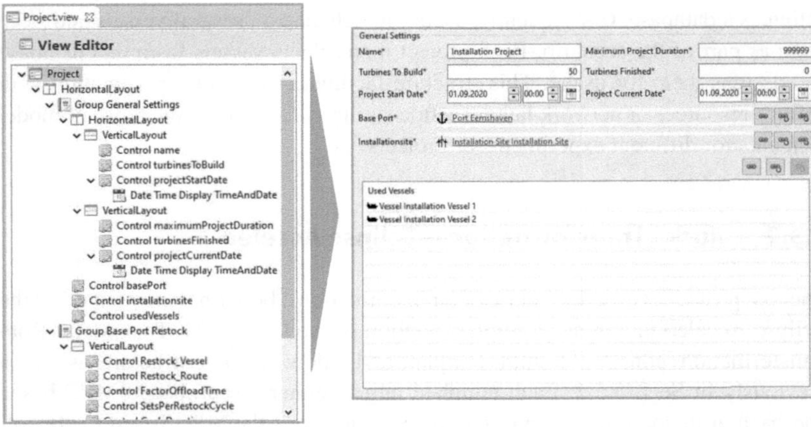

Fig. 6.4 Screenshot of the tool's generated user interface (right) and the EMF Forms generator model (left)

Experimental Setup and Base-Scenario 7

This chapter describes the experimental setup used to evaluate the proposed DSS, the underlying methodology, and its components. It first presents the base-scenario used throughout the evaluation in Sect. 7.1. This base-scenario follows a realistic use-case described in the literature and extends it with additional openly available information. Section 7.2 then presents a short demonstration of how the DSS solves this base-scenario. This demonstration mainly depicts the iterative solution strategy implied by the MPC. Afterward, Sect. 7.3 presents a verification of the MAS to establish that it represents the same processes as the the proposed methodology. Finally, it details the experimental design in Sect. 7.4. This design shows which experiments this thesis covers and what they aim at evaluating.

7.1 Base-Scenario

This section presents the base-scenario used throughout the experiments. If not stated otherwise, all experiments rely on the parametrization given in this section. Thereby, the base-scenario already incorporates parametrizations that resulted from the experiments described later on. Generally, the base-scenario resembles the scenario described in Beinke et al. (2017). It covers a real-world use-case and incorporates values collected throughout several research projects in the area of offshore wind energy conducted at BIBA - Bremer Institut für Produktion und Logistik GmbH at the University of Bremen. Other numbers, e.g., capacities or weights, have been collected from the literature or other openly accessible sources.

© The Author(s), under exclusive license to Springer Fachmedien Wiesbaden GmbH, part of Springer Nature 2026

D. Rippel, *Online Decision Support for Offshore Wind Farm Installations*,
https://doi.org/10.1007/978-3-658-49912-9_7

7.1.1 General Characteristics

The base scenario covers the installation of 50 turbines at a construction site located in Germany's Northern Sea. The scenario locates the base port in Eemshaven, Netherlands, and the components' production ports in Cuxhaven, Germany (towers), and Bremerhaven, Germany (blades and nacelles). The scenario assumes a single installation vessel and an additional HLV for the resupply of components from the production ports to the base port. It assumes that all vessels are located in the base port and are empty in the beginning. Following current literature, this thesis assumes that an installation vessel can transport up to four full component sets on a single trip. If not stated otherwise, construction starts on the first of June 2000 (c.f. Sect. 7.1.5 for details). The scenario assumes the base port to provide one loading bridge/bay always available to the installation vessel and storage capacity for 32 component sets. The initial inventory includes twenty full sets of components, each consisting of one tower, one nacelle and hub, and three blades. These values for the capacity result from the later-described experiment and constitute a realistic upper bound to avoid waiting times due to missing inventory in all optimized instances. In practice, installation projects usually employ a single installation vessel due to their extensive costs. Some experiments covered later-on in this thesis also include additional vessels to test the approaches behavior in such situations. Therefore, scenarios using the default base port storage and capacity levels usually apply a maximum of two vessels. Further vessels induce excessive waiting times due to missing components. Experiments using more than two vessels generally assume a fully stocked base port to avoid these waiting times.

The restock cycle for a total of eight sets takes 312 hours as described later in the respective optimization experiment, or deducible from Beinke et al. (2017). The base-scenario utilizes the average space and weight capacity across several existing vessels to allow a realistic estimation of the transport vessel's capacities. Therefore, the spec-sheets of 22 randomly chosen heavy-lift vessels have been averaged, resulting in an average weight capacity of 438 TEU $\approx$ 8900 t and a usable deck area of approximately 2464 m^2. In terms of the size and weight of components, this thesis relies on the average values for current 10MW turbines given in BVGassociates (2019) and summarized later on in Table 7.1. Moreover, the scenario assumes that two sets of blades can be stacked on top of each other to reduce the required deck area. Regarding the HLV's loading and unloading times, this thesis follows the durations provided Beinke et al. (2017).

Table 7.1 Base-Scenario: Supply Network and Transport Planning

Parameter	Tower	Blade	Nacelle
Base port	Eemshaven (53.4544, 6.8386)		
Installation Site	Northern Sea, Germany (54.5, 6.4)		
Vessel Capacity—Space (HLV)	2646 m^2		
Vessel Capacity—Weight (HLV)	8900 t		
Vessel Average Speed (HLV)	9.5 knots		
Production Port	Cuxhaven	Bremerhaven	Bremerhaven
Latitude	53.8654	53.5432	53.5432
Longitude	8.7260	8.5675	8.5675
Loading Time per Component (HLV)	2 h	8 h	10 h
Unloading Time per Component (HLV)	1.2 h	4.8 h	6 h
Setup Time per Component (HLV)	0	0	0
Weight per Set of Component	600 t	240 t	500 t
Space required per Set of Component	650 m^2	300 m^2	263 m^2

The scenario uses the Haversine formulae to calculate the distance between ports and the installation site using their geolocation coordinates in latitude and longitude to account for the earth's curvature. Assuming a maximum vessel speed of 10 knots (9.5 knots average speed), it then calculates all travel durations by dividing the calculated distance in nautical miles by the average speed. Consequently, all distances and travel durations refer to a straight line" between these locations as an approximation for using actual sea roads.

7.1.2 Operations and Process Chains

In terms of operations and their weather restrictions, this thesis uses the values provided in Table 7.2, which have mainly been obtained from the literature (Oelker et al., 2018). This table additionally lists the sequence of operations used by the optimizer. Thus the scheduled installation operations consist of the sequence: (re-)position in the field, jack-up, install tower, install nacelle, three times install blade, install hub, and finally jack-down. The optimizer directly uses the movement and loading operations specified in this table.

Table 7.2 Base-Scenario: Operation Times and Limits: Installation Scheduling

Operations:				
Index	Operation	Base Duration	Max. Wind	Max. Wave
		[hours]	[m/s]	[m]
1	Install Tower	3	12	99
2	Install Nacelle	3	12	99
3	Install Blade	2	10	99
4	Install Hub	2	12	99
5	Move to Port/Site	4	21	2.5
6	Reposition in Field	1	14	2
7, 8	Jack-up and -down	2	14	1.8
9	Load / Unload (IV)	12	99	99
Aggregated operations used by optimizer:				
1	Install OWT	Consists of: 6, 7, 1, 2, 3, 3, 3, 4, 8		
2	Move to Port	Consists of: 5		
3	Move to Site	Consists of: 5		
4	Load	Consists of: 9		

7.1.3 Estimation of Cost Parameters

The scheduling model uses different types of costs associated with the installation process: Costs for vessels being offshore, costs for sailing operations, and costs for port operations. In reality, these costs depend on several factors and vary strongly from project to project. Nevertheless, the base-scenario used in the evaluations tries to provide realistic estimates for these costs using current literature and available information.

Cost for being offshore: According to literature and the experts' opinions, costs for remaining in the port or being offshore differ by approximately 20% of the charter rate (Rippel et al., 2019d). This percentage includes, e.g., contractual costs, personnel, or energy costs. As described in Chap. 2, current literature states charter rates between approximately 70,000€ and 330,000€ per day per vessel, e.g., in (Dalgic et al., 2013, Meyer, 2014). This thesis assumes an average charter rate of 200,000€ per day, equaling an average cost of approximately 8,333€ per hour, to represent a generalized case. Moreover, it assumes a rounded value of 9,000€ per hour to accommodate the comparably old figures used in this calculation and the expected increase in price (compare Sect. 2.1). Consequently, the additional cost for being offshore amounts to $9,000 \cdot 0.2 = 1,800$€ per hour.

Cost for (un-)loading operations: Next to the technical reason of preventing the optimizer from scheduling unnecessary loading operations, port operations contribute a non-negligible cost factor that should be regarded during the optimization. These port operations require heavy-duty loading gear, like cargo-bridges and towing machines, to transport components from storage to the bay and finally onto the installation vessel. This thesis relies on exemplary renting prices provided online by the Eurogate Container Terminal in Wilhelmshaven, Germany (EUROGATE, 2019). The current price list states hourly rental prices for cargo bridges of 1,325.50€ and 175.50€ for towing machines. Assuming slightly better prices due to the long duration of the installation project, the hourly cost for this equipment amounts to approximately 1,200.00€ per hour at a discount rate of randomly chosen 20%.

Cost for traveling (fuel): A review of current literature shows that basically no viable information on the fuel consumption of jack-up installation vessels exists in literature. Nevertheless, most shipping companies offer spec-sheets for their fleet, and some of these provide an estimate of the fuel consumption per day, as, e.g., provided at Fred. Olsen Windcarrier (2020). These spec-sheets display the fuel consumption at a particular traveling speed over 24 hours. This base-scenario assumes the average consumption of the listed vessels for a normalized vessel speed of 10 knots per hour to obtain a generalized assumption. Normalizing and averaging the values given in Fred. Olsen Windcarrier (2020) results in an hourly consumption of approximately 1.71 tonnes per hour while traveling at 10 knots. This estimation neglects any non-linear effects of the speed on fuel consumption. Regarding the price of vessel fuel, PBT Netherlands BV (2020) lists the current price of vessel fuels per ton for associated ports. Averaging the price for standard low sulfur marine-diesel fuels (VLSFO 0.5%) across the listed 34 ports results in an average price of 297.29€. This value was calculated on 26.06.2020 using the daily price per ton and the current exchange rate from USD to EUR. Current prices for fossil fuels have dropped drastically by approximately 20% between January and May 2020 in Germany (Statistisches Bundesamt Deutschland, 2020) due to the COVID-19 pandemic. Consequently, this base-scenario assumes a cost of approximately 356.75€ per ton ($297.29 \cdot 1.2$) to accommodate this special situation. Multiplying this value with the average fuel consumption described above, the price per hour of traveling amounts to approximately 606.47€. For the sake of an easier calculation and representation, the base-scenario assumes an average cost of 600€ per hour.

Besides the incurring costs for operations, the scheduling formulation uses several penalty terms, which describe bonuses, e.g., for installing an OWT or for finishing operations early. While these penalty terms mainly try to introduce incentives for the optimizer and aim to reduce the formulation's symmetry, too large or too low values can impede the optimization both in terms of the solution quality and the optimization speed.

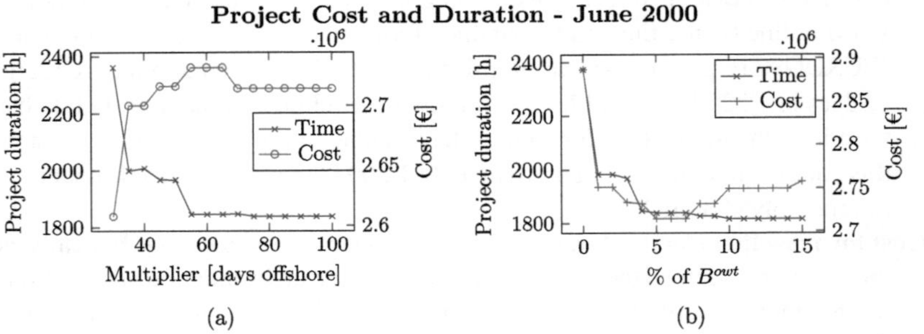

Fig. 7.1 Effect of penalty terms on the cost and duration of the installation project: (a) B^{owt} (b) B^{early} in percent of B^{owt}

Figure 7.1 shows an evaluation of the parameters B^{owt} and B^{early} for the base-scenario. The first figure shows that assigning a bonus of 75 days worth of the costs of being offshore to finishing an OWT shows a good trade-off between scheduling operations that incur waiting times or not. This bonus needs to be higher than the incurring costs for loading, traveling, and the installation ($\geq 26 \cdot C_v^{offshore}$), but should not be too high to render extensive offshore waiting times viable. Similarly, the bonus for finishing an operation early within the planning horizon should be high enough to reduce the scheduling model's symmetry. However, it should not render unnecessary loading or traveling operations viable. The second figure shows that selecting this bonus as approximately 6% of the bonus for finishing an OWT gives good results. Evaluating these parameters with other possible starting dates shows the same overall results. Finally, the inventory penalty B^{inv} can be chosen more or less freely to express the models aim to either keep the inventory full or prefer minimal inventory levels. By default, the base-scenario chooses this parameter as -10 to prefer high inventory levels.

7.1.4 Parameters for the MPC and Optimizer

Considering the MPC's configuration, the base-scenario chooses the sampling step width T as half a week (84 hours) and the number of prediction periods P as two. While higher numbers result in slightly better results, these settings provide a viable trade-off between computational speed and solution quality (c.f. Sect. 8.2.2). Consequently, the scenario generally limits the optimization time per instance to fifteen minutes, if not stated otherwise. This time usually suffices to solve the imposed scheduling problem. While some experiments limit the targeted plan length in hours, the base-scenario generally does not impose a limit on the plan length but instead measures the time required to install all turbines. Some of the experiments described later diverge from this mode.

7.1.5 Weather Scenarios

The proposed DSS provides optimized schedules while considering different weather conditions and their respective forecasts. Accordingly, the experiments presented later cover different scenarios, composed of distinct good, moderate, and bad weather conditions. Given the base-scenario described above, most projects using a single installation vessel take approximately two months (61 days) to finish. Thus, this thesis first analyzed the expected installation durations for each project starting month in 2000 to determine months with distinctive weather conditions. Therefore this thesis applies the already introduced methods to estimate the duration of operations. Table 7.3 shows the results of this analysis. A mean duration of 19 hours would describe perfect weather, i.e., within these two months, installations would not need to delay at any time. High values indicate generally bad conditions, i.e., the conditions induce either frequent or extreme delays. The second row shows the percentage of operation starting hours that induce a delay. Thus, scenarios with high mean durations and low percentages include few but extreme weather conditions, while high percentages indicate generally bad weather. The thesis uses the year 2000 as a baseline to ensure that a maximum of historical data remains accessible,

Table 7.3 Average installation duration for periods of two months in the year 2000

Month	April	May	June	July	August	September	October
Mean Duration [h]	23.5	25.6	27.9	24.4	22.6	57.7	72.5
Hours w. delay [%]	25.8	33.0	37.4	23.4	24.3	52.0	76.0

e.g., for capacity and transport planning. Moreover, comparing the years 2000 until 2006 (last complete dataset), the year 2000 presents the most distinctive conditions within a single year.

The results in Table 7.3 indicate that August provides the best weather conditions, with comparably few delayed starting hours and the lowest overall mean duration. In contrast, October constitutes the scenario with the worst conditions. Nevertheless, experiments showed that selecting October often induces such high delays that the installation projects reach into the next year due to increasingly worse weather conditions in later months. Consequently, this thesis uses September as its "bad conditions" scenario. Finally, this thesis uses June as a moderate weather scenario. Despite its still low mean duration, June shows a high percentage of delayed hours, indicating a high frequency of windows with moderate to bad weather. Figure 7.2 shows the calculated duration for these scenarios if an installation operation would start at a given hour k.

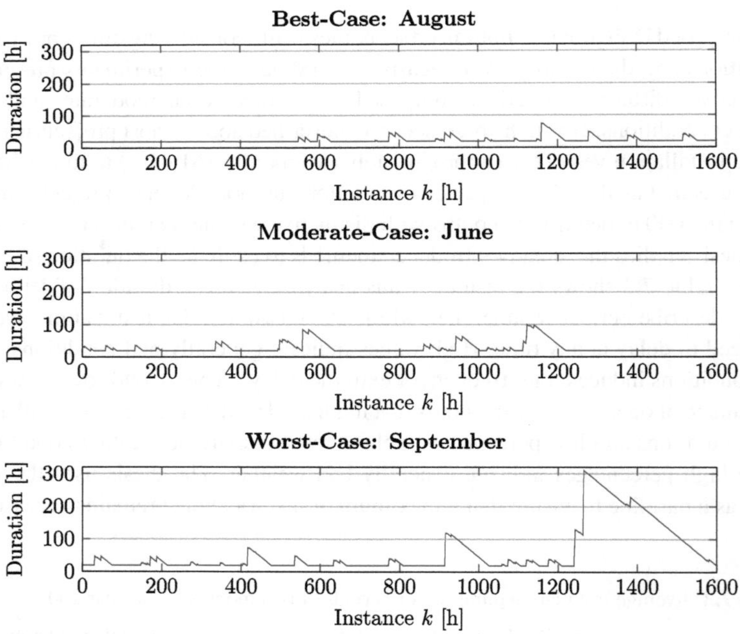

Fig. 7.2 Installation duration for the first 1600 hours of June, August, and September

7.2 Demonstration of Solving the Default Scenario

This section shortly presents an application of the proposed DSS to solve the default scenario presented above. This demonstration serves as a proof of concept that the underlying methodology provides the capabilities for online scheduling under forecast uncertainty and shows the process of simulating an installation project. Therefore, this demonstration first creates an input file for the optimizer using the scenario manager (c.f. Sect. 6.3) using the default scenario. Afterward, it executes the scenario by incrementally obtaining plans and applying the first half week using the simulation component.

Figure 7.3a shows the results of this incremental planning for the first 512 hours of the project, and Fig. 7.3b shows the utilized incremental plan iterations. It shows applied operations or operations within the sampling step width (84 hours) with a black borderline and planned operations with a lighter, gray border. Moreover, each installation operation lists its actual (a) or planned (b) duration. Additionally, the applied plan in Fig. 7.3a shows if an operation diverges from its plan. For example,

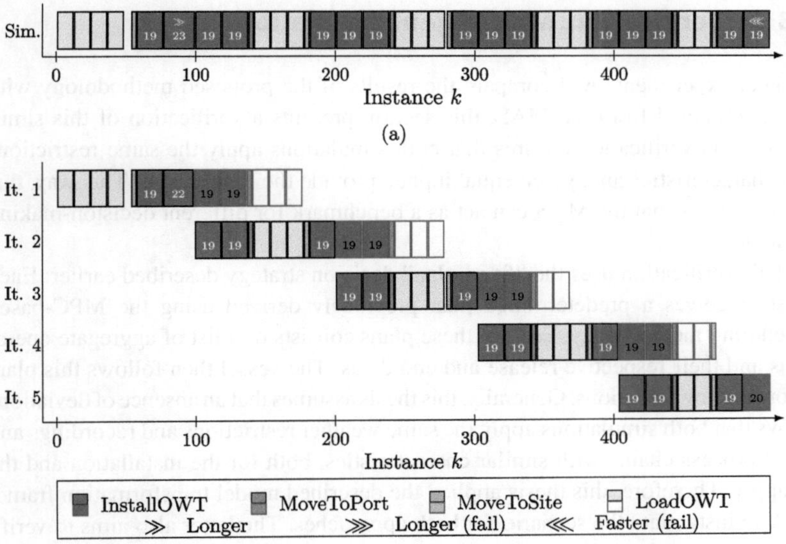

Fig. 7.3 Planing results for the first 512 hours using $P = 2$. (a) Applied operations (b) First five incremental plans

the second installation took longer than planned, while the last operation finished faster.

Both figures only show the first 512 hours of the resulting plan for better readability and an easier comparison. Figure 7.3b shows the optimized plans for the first five MPC iterations (It.). For each iteration, the methodology first obtains forecasts as described in Sect. 5.3.2.2. It then applies the Markov-Chain duration estimator (Sect. 5.3.3.2) to obtain each operation's predicted duration within the prediction horizon. Afterward, it applies the MILP scheduling model to determine the shown plan and applies the first sampling step using its simulation component. The applied plan (Fig. 7.3a) shows good adherence to these plans. Only two operations show deviations due to forecasting uncertainties. The figure shows that both operations constitute the end of the current plan, where uncertainties are highest.

In conclusion, this demonstration shows how the proposed DSS obtains its plans and simulates them. While this demonstration provides a proof of concept, the following section and chapter describe several in-depth evaluations. These aim to determine characteristics and provide benchmarks for the underlying methodology.

7.3 Verification: Multi-Agent Simulation

As later experiments will compare the results of the proposed methodology with results obtained from the MAS, this section presents a verification of this simulation. This verification ensures that both simulations apply the same restrictions and characteristics and, given equal inputs, provide the same result. This way, this thesis ensures that the MAS can act as a benchmark for different decision-making strategies.

This verification uses the "fixed-plan" decision strategy described earlier: Each vessel receives a predetermined plan previously derived using the MPC-based scheduling methodology. Each of these plans consists of a list of aggregate operations and their respective release and end dates. The vessel then follows this plan, recording any deviations. Generally, this thesis assumes that an absence of deviations shows that both simulations apply the same weather restrictions and recordings and equal process chains with similar characteristics, both for the installation and the resupply. Therefore, this thesis applied the described model transformation framework to instantiate the scenario for both approaches. The latter also aims to verify the implemented transformations practically.

The verification covers two stages, as indicated in Table 7.4. The first stage aims to verify the underlying weather assumptions and restrictions. This stage assumes that the port storage already contains all 50 component sets. It allows a sensible use of one to three installation vessels for each of the starting months, resulting in a total of nine experiments. The second stage includes six additional simulation runs, alternating between one and two installation vessels for each of the three selected months while using the base port storage's default settings. This stage aims to verify a similar behavior, considering the base port resupply, and provides further experiments for the overall verification.

Table 7.4 Settings for the verification of the MAS

Parameter	Symbol	Domain
Starting date	–	June, August, September
Installation vessels	V	1, 2 (3)
Initial Storage	X_0^{invP}	20 (50)

For each experiment, the optimizer first created a plan that the agents within the simulation followed. For each operation, AnyLogic recorded when it was released and when it should have been released. Across all experiments, this comparison shows no deviations[1] in the operation's release times or durations for any scenario, resulting in the conclusion that both models represent the same scenario.

7.4 Experimental Design

This section presents the experimental design, considering the design targets and requirements presented in Sect. 2.4.2. While selecting viable methods already satisfied some of these requirements, some require a more detailed evaluation. This need for evaluation mainly concerns the requirements **-R1-** solution quality, **-R2-** real-time capability, **-R3-** support for planning tasks on higher levels, **-R5-** integration of weather forecasts, and **-R7-** aggregation of historical data for higher-level tasks. While some of these target the DSS as a whole, the latter two specifically target its components.

Therefore, the overall experimental design distinguishes between three areas: First, the evaluation of components to determine viable approaches or parameteri-

[1] In some cases, the log files show minor deviations considering the release dates of one hour. On closer examination, these deviations result from rounding errors during the log file generation due to the higher internal time resolution of AnyLogic. Within the simulation, the release times and durations match the originals.

zations. Second, the evaluation of the overall scheduling methodology in terms of its performance. Finally, the third area targets evaluations of offline applications. This area aims to determine the methodology's viability for such higher-level planning tasks and serves as a proof-of-concept. Figure 7.4 shows the experiments covered by these three areas in more detail and lists their respective sections in this thesis.

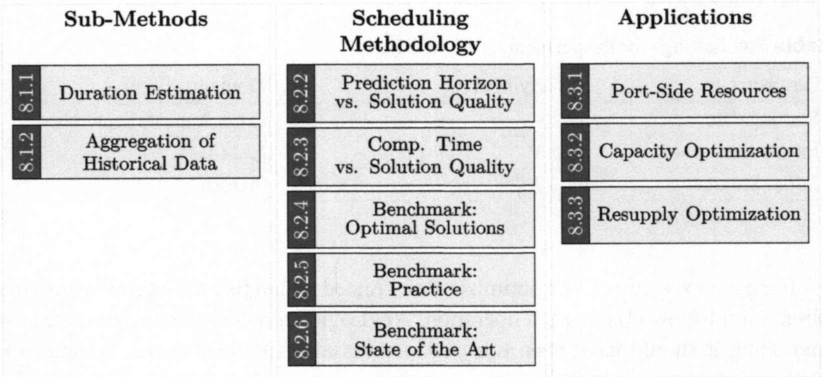

Fig. 7.4 Experiments by area. Vertical markers reference the appropriate sections

The first area covers three different experiments. The integration of forecasting uncertainties into the model and an appropriate duration estimation for operations constitute major development goals for the proposed DSS. Besides proposing a way to integrate these uncertainties into the model, this thesis proposes two different estimators to discretize these uncertainties. The respective experiments aim to evaluate both estimators in terms of their accuracy and select one of them for future experiments. The second experiment aims at determining the influence of different choices for windows of historical weather information on higher-level planning tasks. Again, this thesis proposes two possible approaches to preprocess historical data.

The second area evaluates the underlying methodology regarding its computational speed and solution quality. Therefore, this thesis first presents two experiments to determine the influence of different parameterizations for the planning horizon and the number of prediction periods on the solution quality. Additionally, it evaluates the solution quality under different time constraints by limiting the optimization time. Afterward, this thesis presents various benchmark scenarios. These include comparing the proposed methodology's results with an optimal plan that does not rely on weather forecasts. Furthermore, this thesis presents two experiments using the described MAS to benchmark a simple priority-based decision strategy and a forecasting-based strategy that mimics the current state of practice. Finally, this area includes a benchmark to compare the methodology with robust or online-control optimizations, which can be considered the current state of the art. The majority of these experiments use the plan or project duration and the incurred costs as primary quality criteria. The experiments recalculate the costs using the generated plan to omit the optimizer's penalty terms and obtain more realistic cost estimations.

The third area presents experiments that evaluate the methodology's viability for higher-level planning tasks as part of the DSS. Therefore, it presents the application of the methodology and its components to three different examples. First, the evaluation of port-side resources, i.e., an estimation of the influence of available loading bays for an installation project. Second, the optimization of the base-port storage capacity, and third, the optimization of resupply cycles. Particularly the last evaluation applies and combines the majority of models and methods presented in this thesis.

Experimental Results

8

This chapter presents the results of various experiments regarding the three areas stated in the previous chapter. Therefore, this chapter first presents the experiments relating to the selection and parameterization of components in Sect. 8.1. Afterward, it presents the evaluation of the underlying MPC-based scheduling methodology (Sect. 8.2) and the defined application examples (Sect. 8.3).

8.1 Evaluation of Components

The DSS and its underlying methodology include several components that influence its behavior. This section shortly evaluates different method selections or parameterizations for these components. It aims at determining settings that this thesis uses in the remaining experiments. It includes selecting the duration estimator in Sect. 8.1.1 and an evaluation of how to treat historical weather recordings for offline tasks in Sect. 8.1.2.

8.1.1 Evaluation: Duration Estimators (Scheduling)

This section shortly summarizes the results published in Rippel et al. (2019b). It evaluates the estimation accuracy of both estimators presented in Sect. 5.3.3.

This experiment applies both estimators to the months April to October 2000, each with a prediction horizon of one to four weeks. The sliding-window estimator thereby relies on the parameter ω to determine viable time windows. For each scenario, the experiment optimizes this parameter by repeating the estimation with

© The Author(s), under exclusive license to Springer Fachmedien Wiesbaden GmbH, part of Springer Nature 2026

D. Rippel, *Online Decision Support for Offshore Wind Farm Installations*, https://doi.org/10.1007/978-3-658-49912-9_8

different settings in steps of 0.05. This evaluation uses the best results with the lowest prediction error for comparison. It has to be noted that such an optimization is only possible when working with historical data. In practice, this estimator requires a best-guess, possibly resulting in weaker solutions. In contrast, the Markov-Chain estimator does not rely on any variable parameters. This experiment uses the standard deviation between the actual duration and each estimator's expected duration as quality criteria. The standard deviation provides a viable representation of the general prediction error. The last table column shows the mean values across the year.

Table 8.1 shows the results of these experiments. The top compartment shows the actual duration of installation operations in terms of their minimum, mean, and maximum duration for each month as a reference. The lower compartments depict the standard deviation in hours for both estimators by month and prediction horizon. The results show an increase in the estimation error with increasing prediction horizons. While this error remains comparably low for both estimators for horizons up to two weeks, it increases strongly for the sliding window estimator beyond this point. In contrast, the Markov-Chain estimator's errors remain comparably low, with an average standard deviation of approximately 15 hours.

Table 8.1 Comparison of duration estimators (Rippel et al., 2019b)

2000	Apr.	May	Jun.	Jul.	Aug.	Sep.	Oct.	Mean
Dur. Min	19.0	19.0	19.0	19.0	19.0	19.0	19.0	19.00
Dur. Mean	23.0	23.0	28.0	30.0	20.0	26.0	31.0	25.86
Dur. Max	69.0	81.0	85.0	101.0	41.0	74.0	130.0	83.00
Standard Deviation: Sliding Window Estimator in hours								
1 Week	0.22	1.77	1.41	2.21	0.00	0.57	0.86	1.01
2 Weeks	2.16	1.25	3.29	3.67	0.00	3.11	2.10	2.23
3 Weeks	2.10	3.07	20.81	20.15	0.55	10.18	15.93	10.40
4 Weeks	35.82	77.29	31.79	17.45	1.79	94.42	335.87	84.92
Standard Deviation: Markov-Chain Estimator in hours								
1 Week	0.26	0.67	2.56	0.58	0.00	1.47	0.77	0.90
2 Weeks	1.07	0.53	1.99	2.28	0.08	2.86	6.30	2.16
3 Weeks	1.14	0.95	4.06	12.73	0.29	5.02	5.59	4.25
4 Weeks	1.08	3.07	6.78	11.02	1.26	5.26	77.85	15.19

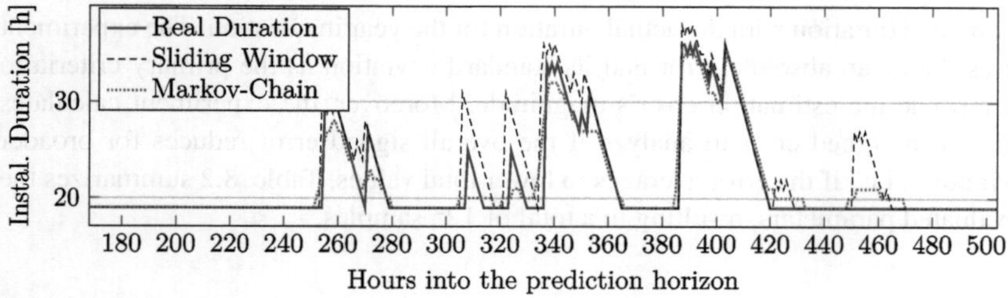

Fig. 8.1 Duration estimation for weeks two and three in April 2000—adapted from Rippel et al. (2019b)

Figure 8.1 shows an example of the estimated durations for the second and third weeks of April. This figure shows the general tendency of both estimators. While the sliding window estimator tends to overestimate the duration, the Markov-Chain estimator tends to underestimate the duration slightly. In the context of this thesis, the Markov-Chain estimator constitutes the more viable alternative. First, it provides more accurate estimations, as shown in Table 8.1. Second, its tendency to underestimate might reduce the number of recourse actions in cases where a possible delay would still fit into the generated plan. In contrast, an overestimation would always result in a recourse action. In conclusion, this experiment allows drawing the following conclusion:

-1- The results show that both estimators provide viable results for the relevant prediction horizons of one, two, and three weeks. Nevertheless, the Markov-Chain estimator provides slightly more reliable results for the planning.

8.1.2 Evaluation: Duration Estimation Using Historical Data (Capacity)

The capacity optimization presented earlier relies on an estimation of operation durations using historical data. Both presented approaches use a window, determined by the number of included past years N^{years}, and, in the case of the aggregation-based method, the length of the resupply cycle RES^{frq}. This section summarizes the results presented in a previous study (Rippel et al., 2020a).

This experiment aims to determine the influence of these parameters on the quality of the estimation. Therefore, this experiment constructs the mean duration of operations for the windows described in Table 8.2. Afterward, it compares the estimated duration with the actual duration for the year in question. The experiment uses the mean absolute error and its standard deviation as the primary criteria to determine the estimation error's magnitude. Moreover, the experiment calculates the mean signed error to analyze if the overall signed error reduces for broader windows, i.e., if the error averages to lower total values. Table 8.2 summarizes the evaluated parameters, resulting in a total of 125 samples.

Table 8.2 Settings for the duration estimation using historical data (Rippel et al., 2020a)

Parameter	Notation	Domain
Year to evaluate		1980, 1985, 1990, 1995, 2000
Past Years to regard	N^{years}	2, 5, 10, 15, 20
Window Sizes (hours)	RES^{frq}	3, 24, 168, 336, 744

Figure 8.2 shows the results of these experiments aggregated by the number of past years (Fig. 8.2a) used and the resupply cycle (Fig. 8.2b). Generally, the results show relatively low mean absolute errors of less than six hours throughout the experiments. Moreover, the low mean unsigned error shows that the actual errors average themselves out quite well.

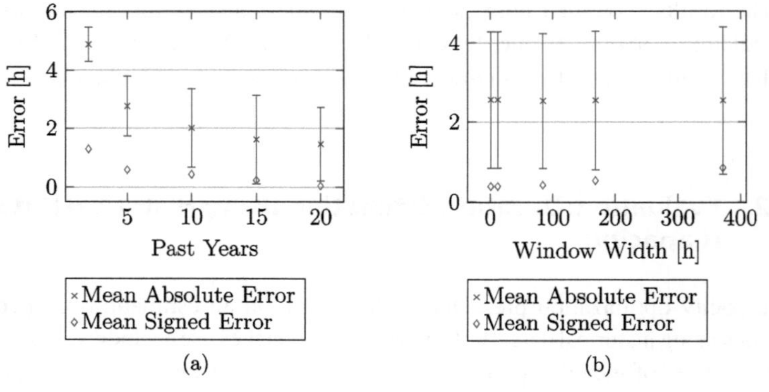

Fig. 8.2 Evaluation of the window size using historical weather data—adapted from Rippel et al. (2020a) © 2020 SCS

The results show no significant impact of the window's width on the prediction accuracy. In contrast, the number of past years included in the prediction decreases both errors. Nevertheless, the results suggest a saturation, starting at approximately ten years of historical data. The current results indicate that including more than 20 years might not yield significantly better results. Similarly, the mean signed error decreases with the inclusion of additional years. In contrast to the initial hypothesis, a broader window slightly increases the signed error, indicating an increasing tendency to overestimate the duration. While the proposed methods cannot freely choose the window's width, this thesis employs $N^{years} = 20$ for all further experiments. The results allow drawing the following conclusions:

-2- The width of the window (length of resupply cycles) does not have a strong influence on the aggregation quality.
-3- The depth of the window (number of past years considered) shows a saturation, starting at including approximately 10–20 years of historical data.
-4- The results show comparably low mean errors, indicating this approach's viability to estimate durations based on historical data.

8.2 Evaluation of the MPC-Based Scheduling Methodology

This section aims to evaluate the proposed scheduling methodology in terms of its logistic and computational performance. Therefore, it first introduces which soft- and hardware was used during the experiments (Sect. 8.2.1). Afterward, it presents two experiments the determine the quality of solutions (plans) for different prediction horizons in Sect. 8.2.2, and under varying computational limitations, in Sect. 8.2.3. Afterward, it presents several benchmarks, comparing the proposed methodology's results with other solutions. These benchmarks cover optimal schedules, which do not rely on forecasts (Sect. 8.2.4), state of practice approaches implemented as part of the AnyLogic MAS (Sect. 8.2.5), and two variants of the proposed methodology to represent state of the art online scheduling approaches (Sect. 8.2.6).

8.2.1 Summary of Used Hard- and Software

This thesis uses an implementation of all the noted methods in Matlab 2018b. Furthermore, it uses CPLEX 12.10 to solve the noted optimization problems using an exact branch-and-bound algorithm. Some of the noted benchmarks use AnyLogic 8.5.2 to execute the MAS. This MAS model also includes a Java-based implementation of the weather forecast generation and the Markov-Chain duration estimator to ensure that both approaches can access the same information. All experiments have been executed on a common desktop PC with an AMD Ryzen 9 3900X 12-Core processor (24 logical units), 32GB RAM, and a 1TB SSD hard-drive. This setup provides CPLEX with a high number of parallel threads (max. 24) to speed up the optimizations.

8.2.2 Evaluation: Prediction Horizon vs. Solution Quality

This experiment evaluates the solution quality and run-time behavior of the proposed approach for different sampling step widths T and different numbers of prediction periods P. Therefore, the experiment includes executions of the approach for each of the three months, using the settings in Table 8.3. The experiment uses the stagnation optimization mode described earlier, with a time limit of 3600 seconds (one hour) until iteration 4 and 7200 seconds (two hours) for all further iterations.

Table 8.3 Settings of the experiment to determine the algorithms run-time behavior

Parameter	Symbol	Domain
Starting date	–	June, August, September
Planning horizon	T	84, 168
Prediction periods	P	1, 2, 3
Installation vessels	V	1, 2, 3
Initial inventory	X_0^{invP}	50
Port capacity	CAP^{port}	50
Loading Bays	$BAYS$	1

The experiments assume full storage at the base port to omit the influence of possibly insufficient resupply. Moreover, it assumes that the port only offers a single loading bay, regardless of how many installation vessels it applies. The experiments use the installation project's cost per installed OWT and the project duration as quality criteria. The ratio between optimal and sub-optimal and the mean computational time per plan as performance criteria. The cost per installed turbine thereby shows more distinctive differences due to the generally lower values. It allows calculating the overall project cost by multiplying this value times 50.

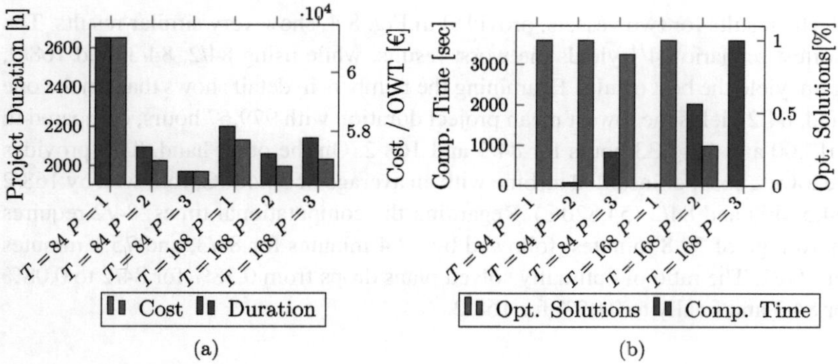

Fig. 8.3 Mean results for $V = 1$ and all months

Figure 8.3 shows the mean results for a single installation vessel in plan quality (a) and algorithm performance (b). Regarding the plan quality, the figure shows that using a short planning horizon $T = 84$ and a single prediction period results in the worst plans with very high project durations and costs, indicating that the model cannot schedule operations efficiently with such a short horizon. Moreover, the graph shows a decrease in both values for increasing prediction periods. This behavior results from the duration of installation cycles, which depends on the vessel's capacity. The prediction horizon needs to allow scheduling a total of four loading and installation operations, amounting to a minimum horizon of 124 hours plus travel times. The figure shows similar behavior for the longer planning horizon $T = 168$. Increasing the number of prediction periods from one to two strongly reduces both the project duration and the costs. Nevertheless, an increase to three periods for the longer planning horizon results in slightly worse plans. This behavior results from the fact that the optimizer could not obtain optimal plans for such large scenarios in time, as given in Fig. 8.3b. This figure shows, that the optimizer achieves optimal solutions for all plans using $T = 84$, $P = 1$ (84/1) and $P = 2$ (84/2), and for $T = 168$, $P = 1$ (168/1). This ratio quickly declines for larger instances. Similarly, the optimizer obtains plans for smaller instances quickly. On average, the optimizations take less than a minute for (84/1) and (168/1), approximately 1.2 minutes for (84/2), 14 minutes for 84/3, and 33.8 minutes for (168/2). Sub-optimal plans, finishing due to stagnation (168/3), attribute either 3600 or 7200 seconds to the average, depending on the number of iterations required.

The results for two vessels, provided in Fig. 8.4, show very similar results. The shortest scenario 84/1 yields the worst results, while using 84/2, 84/3, and 168/2, again, yield the best results. Examining the numbers in detail shows that, on the one hand, 84/2 yields the lowest mean project duration with 979.67 hours, compared to 1017.00 and 1011.33 hours for 84/3 and 168/2. On the other hand, 84/3 provides the lowest cost per installed turbine with an average of 54.460€, followed by 168/2 (54,520€) and 84/2 (54,820€). Regarding the computational times, 84/2 requires an average of 39.8 minutes, followed by 52.4 minutes for 84/3, and 75.1 minutes for 168/2. The ratio of optimally solved plans drops from 0.38% for 84/2 to 0.08% for 84/3 and finally to 0.05% for 168/2.

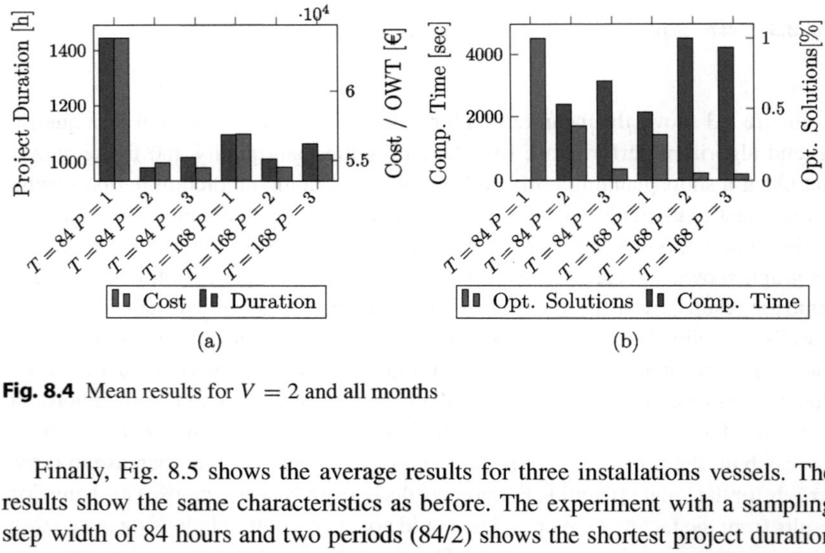

Fig. 8.4 Mean results for $V = 2$ and all months

Finally, Fig. 8.5 shows the average results for three installations vessels. The results show the same characteristics as before. The experiment with a sampling step width of 84 hours and two periods (84/2) shows the shortest project duration with 727.3 hours, followed by 84/3 with 735.7 hours and 168/3 with 763.7 hours. As before, 84/2 shows comparably low costs per OWT of 55,584€ but only ranks fourth in this experiment. 84/3 achieves the lowest costs with 54,684€, followed by 168/3 with 55,236€, and 168/2 with 55,312€ per turbine. This experiment shows similar trends as before, considering the optimization time and the optimality ratio. In contrast, 84/3, 168/2, and 168/3 did not solve any instances to optimality.

In conclusion, these results show that the MPC scheme's application shows distinct advantages over a simple rolling horizon planning ($P > 1$ vs. $P = 1$). Comparing settings with the same total prediction horizon (84/2 and 168/1), both settings obtain optimal results for all plans employing a single vessel using very low

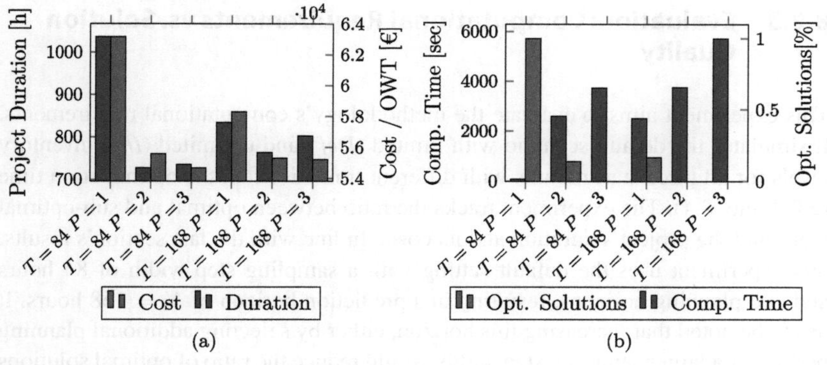

Fig. 8.5 Mean results for $V = 3$ and all months

computational times. Nonetheless, the MPC-based approach shows better results in terms of the plan quality than a simple rolling horizon planning. Moreover, the shorter planning horizon of $T = 84$ provides generally better results. These results indicate a higher adaptability to changing weather conditions and lower uncertainties introduced by forecasts. In terms of this thesis, most experiments use 84/2 as a trade-off between quality and performance. While using more planning periods results in higher quality solutions, it highly increases the algorithm's run-time and introduces higher numbers of sub-optimal plans into the overall planning. This section's results fall in line with previous results published in Rippel et al. (2019d) and Rippel et al. (2019b). Those experiments used a reduced version of the MILP scheduling model but yielded comparable results. The results of those experiments and presented in this section allow drawing the following conclusions:

- **-5-** The use of the MPC-based approach with more planning periods provides better plans than reactive online planning. The results for 84/2 and 168/1 use similar prediction horizons N, whereas 84/2 consistently provides better results.
- **-6-** Lower sampling step widths generally show better results, as they allow the approach to mitigate increasing forecast uncertainties.
- **-7-** The approach requires a minimum prediction period N of one week to obtain viable results.
- **-8-** With currently available hardware, 84/2 or 84/3 perform best and provide a viable trade-off between logistic and computational performance.

8.2.3 Evaluation: Computational Requirements vs. Solution Quality

This experiment aims to evaluate the methodology's computational requirements. It simulates the default scenario with limited (*Def*) and unlimited (*Inf*) inventory levels for all three start months with different limitations on the optimization time (c.f. Table 8.4). The experiment tracks the ratio between optimal and sub-optimal plans and the project's duration and its costs. In line with the last section's results, this experiment uses the default setting with a sampling step width of 84 hours and two planning horizons, resulting in a prediction horizon of $N = 168$ hours. It has to be noted that increasing this horizon, either by selecting additional planning periods or a larger sampling step width, would reduce the ratio of optimal solutions within each simulation but probably result in a better logistical performance. The results presented in this section also support this assumption, as adding a second vessel, and, thus, close to doubling the problem size, heavily impacts this ratio. Nevertheless, previous results suggest that this setting provides reasonable results and serves as a base-line for this experiment. Moreover, the results presented in this section depend on the used hard- and software. As stated before, this thesis uses a standard desktop computer with an AMD Ryzen 9 3900X 12-Core processor (usually 23 active threads) and 32GB of memory for the experiments. Slower of faster computers, especially computers with different numbers of cores or clock speeds, result in different results. Previous results of a similar experiment described in Rippel et al. (2019b) support this assumption. The stated experiment used a weaker computer and achieved comparable results with slightly higher computational times.

Tables 8.5 and 8.6 show the results of the newly conducted experiment. The first column describes the setting given as the inventory setting, followed by the maximum optimization time in seconds, the month, and finally, the number of used installation vessels. Afterward, these tables show the most relevant performance criteria, i.e., the project's duration and costs. Finally, the tables provide the ratio

Table 8.4 Settings of the experiment to determine computational requirements

Parameter	Symbol	Domain
Starting date	–	June, August, September
Installation vessels	V	1, 2
Initial inventory	X_0^{invP}	32 (Def), 50 (Inf)
Port capacity	CAP^{port}	20 (Def), 50 (Inf)
Optimization time [sec]	–	60, 120, 300, 600, 1800, 3600

Table 8.5 Results for the infinite inventory setting at different computational times

Setting	Duration [h]	Cost [€]	$ratio_{opt}$	Opt. Plans	Sub. Plans
Inf-June-60-V1	1780	2,841,600	0.76	16	5
Inf-June-120-V1	1780	2,841,600	0.95	20	1
Inf-June-300-V1	1780	2,841,600	0.95	20	1
Inf-June-600-V1	1780	2,841,600	1.00	21	0
Inf-June-1800-V1	1780	2,841,600	1.00	21	0
Inf-June-3600-V1	1780	2,841,600	1.00	21	0
Inf-Aug.-60-V1	1814	2,734,800	0.95	21	1
Inf-Aug.-120-V1	1795	2,725,800	1.00	21	0
Inf-Aug.-300-V1	1795	2,725,800	1.00	21	0
Inf-Aug.-600-V1	1795	2,725,800	1.00	21	0
Inf-Aug.-1800-V1	1795	2,725,800	1.00	21	0
Inf-Aug.-3600-V1	1795	2,725,800	1.00	21	0
Inf-Sep.-60-V1	2704	2,980,200	0.97	36	1
Inf-Sep.-120-V1	2704	2,980,200	0.97	36	1
Inf-Sep.-300-V1	2704	2,980,200	0.97	36	1
Inf-Sep.-600-V1	2704	2,980,200	1.00	37	0
Inf-Sep.-1800-V1	2704	2,980,200	1.00	37	0
Inf-Sep.-3600-V1	2704	2,980,200	1.00	37	0
Inf-June-60-V2	1080	2,865,600	0.07	1	13
Inf-June-120-V2	1063	2,822,400	0.07	1	13
Inf-June-300-V2	1027	2,805,000	0.08	1	12
Inf-June-600-V2	964	2,793,000	0.00	0	12
Inf-June-1800-V2	963	2,789,400	0.08	1	11
Inf-June-3600-V2	963	2,793,000	0.42	5	7
Inf-Aug.-60-V2	932	2,712,000	0.00	0	11
Inf-Aug.-120-V2	978	2,725,200	0.08	1	11
Inf-Aug.-300-V2	909	2,695,800	0.00	0	11
Inf-Aug.-600-V2	881	2,676,600	0.18	2	9
Inf-Aug.-1800-V2	879	2,695,800	0.09	1	10
Inf-Aug.-3600-V2	897	2,721,000	0.09	1	10
Inf-Sep.-60-V2	1070	2,782,200	0.07	1	13
Inf-Sep.-120-V2	1069	2,720,400	0.13	2	13
Inf-Sep.-300-V2	1070	2,760,000	0.21	3	11
Inf-Sep.-600-V2	1070	2,720,400	0.14	2	12
Inf-Sep.-1800-V2	1051	2,737,800	0.36	5	9
Inf-Sep.-3600-V2	1070	2,730,600	0.50	7	7

Table 8.6 Results for the default scenario at different computational times

Setting	Duration [h]	Cost [€]	$ratio_{opt}$	Opt. Plans	Sub. Plans
Def-June-60-V1	1838	2,731,800	0.68	15	7
Def-June-120-V1	1840	2,714,400	0.86	19	3
Def-June-300-V1	1840	2,714,400	1.00	22	0
Def-June-600-V1	1840	2,714,400	1.00	22	0
Def-June-1800-V1	1840	2,714,400	1.00	22	0
Def-June-3600-V1	1840	2,714,400	1.00	22	0
Def-Aug.-60-V1	1814	2,725,800	0.75	15	5
Def-Aug.-120-V1	1814	2,725,800	0.95	19	1
Def-Aug.-300-V1	1814	2,725,800	1.00	20	0
Def-Aug.-600-V1	1814	2,725,800	1.00	20	0
Def-Aug.-1800-V1	1814	2,725,800	1.00	20	0
Def-Aug.-3600-V1	1814	2,725,800	1.00	20	0
Def-Sep.-60-V1	2719	2,915,400	0.94	31	2
Def-Sep.-120-V1	2719	2,915,400	0.97	32	1
Def-Sep.-300-V1	2719	2,915,400	1.00	33	0
Def-Sep.-600-V1	2719	2,915,400	1.00	33	0
Def-Sep.-1800-V1	2719	2,915,400	1.00	33	0
Def-Sep.-3600-V1	2719	2,915,400	1.00	33	0
Def-June-60-V2	1410	2,934,600	0.22	4	14
Def-June-120-V2	1386	2,891,400	0.25	5	15
Def-June-300-V2	1386	2,847,600	0.28	5	13
Def-June-600-V2	1390	2,770,800	0.47	9	10
Def-June-1800-V2	1408	2,737,200	0.58	11	8
Def-June-3600-V2	1404	2,772,000	0.47	9	10
Def-Aug.-60-V2	1390	2,732,400	0.26	5	14
Def-Aug.-120-V2	1376	2,736,000	0.28	5	13
Def-Aug.-300-V2	1393	2,732,400	0.37	7	12
Def-Aug.-600-V2	1391	2,758,800	0.42	8	11
Def-Aug.-1800-V2	1392	2,711,400	0.44	8	10
Def-Aug.-3600-V2	1392	2,716,800	0.44	8	10
Def-Sep.-60-V2	1822	2,963,400	0.26	6	17
Def-Sep.-120-V2	1822	3,082,200	0.39	9	14
Def-Sep.-300-V2	1822	2,946,000	0.33	8	16
Def-Sep.-600-V2	1822	2,939,400	0.52	13	12
Def-Sep.-1800-V2	1822	2,934,000	0.63	15	9
Def-Sep.-3600-V2	1822	2,928,600	0.54	13	11

of optimal plans and the total numbers of optimal and sub-optimal plans for each simulation run.

The results in these tables show that the optimizer obtains the same plan for all scenarios with a single installation vessel with at least 2 minutes of optimization time (120 seconds). While some of these simulations still contain sub-optimal plans up to a limit of 10 minutes (600 sec), sub-optimal plans do not impact the result, neither regarding the project duration nor regarding the costs. Consequently, this thesis assumes that the optimizer already determined the optimal solution but could not prove its optimality in the given time.

The scenarios using two vessels show a different behavior between the two inventory settings. While the infinite inventory setting requires a lower number of overall plans, the number of sub-optimal plans remains comparably stable for most of the time limits. In contrast, the default inventory setting requires a higher total number of plans, while the number of sub-optimal plans continuously declines with increasing limits. This behavior tracks back to out-of-inventory or very-low-inventory events during the default scenario, resulting in the extended project duration. Consequently, these simulations include periods in which the optimizer cannot schedule any, or only very few, operations as the base port has no components. Such periods simplify the optimization problem and result in additional optimal instances. Apart from these differences, both scenarios show relatively stable values for the project's duration and cost. The scenarios show a slight trend towards shorter projects with an increasing time limit. The same trend shows in the project's costs. Nevertheless, both values sometimes increase with an increased limit, e.g., the duration from Def-Aug.-300-V2 to Def-Aug.-600-V2. In such cases, the optimizer generally made small adjustments to its plans, sometimes resulting in miscalculated operation durations or impacting the plan later on. The plans two and three for September depicted in the next subsection's Fig. 8.7 shows an example of such a modification. In that case, the optimizer preempts an operation to the plan's beginning, which delays later installation cycles and leads to a loss of overall performance.

Figure 8.6 depicts the described trends graphically. Each plot's x-axis shows the different optimization limits on the logarithmic scale. The three left plots show the default inventory setting results, while the right plots show the results for the unlimited inventory setting. The plots show the ratio of optimal plans, the project's duration, and its costs from top to bottom. In general, this experiment allows drawing the following conclusions:

-9- The proposed methodology can solve all plans within ten minutes or less per plan for a default setting with a single installation vessel and a prediction horizon of one week, given the described hardware.

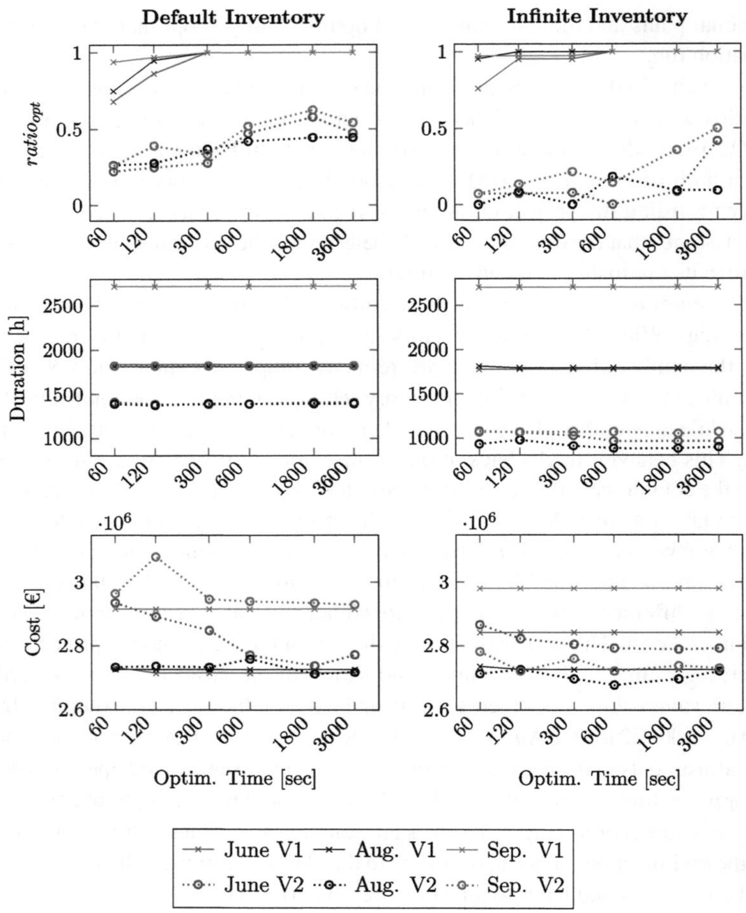

Fig. 8.6 Results computational time and inventory setting (x-axis in log-scale)

-10- Larger problem instances (e.g., for two vessels) cannot be solved optimally with the given hardware. Nevertheless, increasing optimization times only result in slightly better solutions, indicating close to optimal solutions already with lower times.

-11- Overall, even with only one minute, the proposed methodology achieves high-quality results with comparably low differences to the best-found solutions.

8.2.4 Benchmark: Comparison to Optimal Solutions

This experiment aims to estimate the efficiency of the proposed methodology compared to an optimal schedule. Therefore, it compares the schedules of the experiment in Sect. 8.2.2 with optimal plans. Unfortunately, the size of such installation projects prevents solving a complete project to optimality. Consequently, the optimal solution only covers the first four weeks of each project for a single vessel. Therefore, the experiment obtained an optimal schedule for 672 hours (four weeks) plus an extra 100 hours to retain comparability between the receding horizon scheduling and the optimal schedule. Afterward, it pruned the resulting plans back to four weeks (672 hours). This process allows both results to contain extra loading or movement operations and ensures comparable results. The optimal schedules apply the same methods and models presented earlier, e.g., to calculate the duration of operations and the scheduling. In contrast, the optimal schedules do not include forecast uncertainties and use the actual duration of operations. After pruning both plans to four weeks, the experiment calculates the relevant key performance indicators given in Table 8.7 to enable a direct comparison. Finally, even while considering only four weeks, the optimizer could not prove the optimality of the resulting optimal plans. Nevertheless, after several consecutive runs, each of one hour, the plans show a comparably small optimality gap of 2.55% (June), 2.19% (August), and 2.23% (September). These gaps depict the difference between the best incumbent (solution) and the best possible relaxed (non-integer) solution found by the optimizer so far.

Table 8.7 Results of the comparison with optimal schedules

Month	Plan	Length [h]	Cost [€]	OWTs	Cost/ Turbine [€]		Time Offshore/ Turbine [h]	
June	Opt.	656	972,600	17	57,211.76		21.82	
June	$P = 2$	656	1,113,000	18	61,833.33	(+8.0%)	24.50	(+12.3%)
June	$P = 3$	656	1,113,000	18	61,833.33	(+8.0%)	24.50	(+12.3%)
Aug.	Opt	668	1,081,200	20	54,060.00		20.90	
Aug.	$P = 2$	667	1,145,400	20	57,270.00	(+5.9%)	22.55	(+7.9%)
Aug.	$P = 3$	663	1,115,400	20	55,770.00	(+3.2%)	21.85	(+4.5%)
Sep.	Opt	665	958,200	17	56,364.71		21.12	
Sep.	$P = 2$	662	1,064,400	18	59,133.33	(+4.9%)	23.22	(+9.9%)
Sep.	$P = 3$	663	1,135,800	18	63,100.00	(+11.9%)	24.83	(+17.6%)

Table 8.7 lists the performance indicators for the pruned plans. It lists the total duration of the plan, its costs, and the number of installed OWT. As the latter differs between plans, the table also lists relative costs and offshore time per turbine. These relative values provide a better comparison of the plans.

The results show a direct influence of the weather forecasts and the respective conditions on the generated plans' quality. As a month with close to perfect weather conditions, August shows the lowest difference from the optimal plan with average gaps of 4.6% for the costs and 6.2% for the offshore time. Moreover, using a higher number of prediction periods provides better results and benefits from the stable weather conditions. In June, as a month with moderate weather conditions, both plans perform equally well. Nevertheless, the general gap to the optimal plan increases to 8% for the costs and 12.3% for the offshore time. As a month with highly unstable weather, September shows higher average gaps of 8.4% for the costs and 13.8% for the offshore times. Moreover, this month favors the plan with fewer prediction horizons as the optimizer can work with shorter, more stable weather forecasts. Using only two planning horizons cuts the gap approximately in half, compared to using three horizons. This behavior supports using a lower number of prediction periods: Under normal or bad weather conditions, this setting performs equally or better (approximately by 7%). In cases of very good weather, this setting performs slightly worse by approximately 3%.

The results show a general trend of the MPC-based scheduling methodology to install more turbines in a shorter time than the optimal plan indicates. As described, this tendency results in higher costs for each turbine. Nevertheless, the results show that the proposed methodology installed one more turbine in a comparable, even slightly shorter, period in the two non-perfect-weather months (c.f. columns *Length* and *OWTs*). Such behavior could significantly reduce the fixed costs of an installation project by reducing its overall duration. Nevertheless, this thesis does not include fixed costs and considers a further investigation of this effect out of scope.

Figure 8.7 shows a direct comparison of the resulting plans. For each month, Plan 1 depicts the optimal plan, while Plans 2 and 3 show the plans resulting from the MPC-based methodology applying $T = 84$, $P = 2$, and $P = 3$, respectively. Below the plans, each plot shows the actual, non-forecast duration of an installation operation if it started at a given hour to give an impression on the plans' quality. The plots for June show that both $P = 2$ and $P = 3$ resulted in the same overall plan. This plan shows three forecast induced-plan failures, whereby one operation took longer than expected, and two operations turned out shorter than planned. These operations indicate that the optimizer tried to fit as many operations as possible, even at slightly higher costs. In contrast, the optimal schedule only includes a single operation with an increased operation time within the first installation cycle. Comparing the three

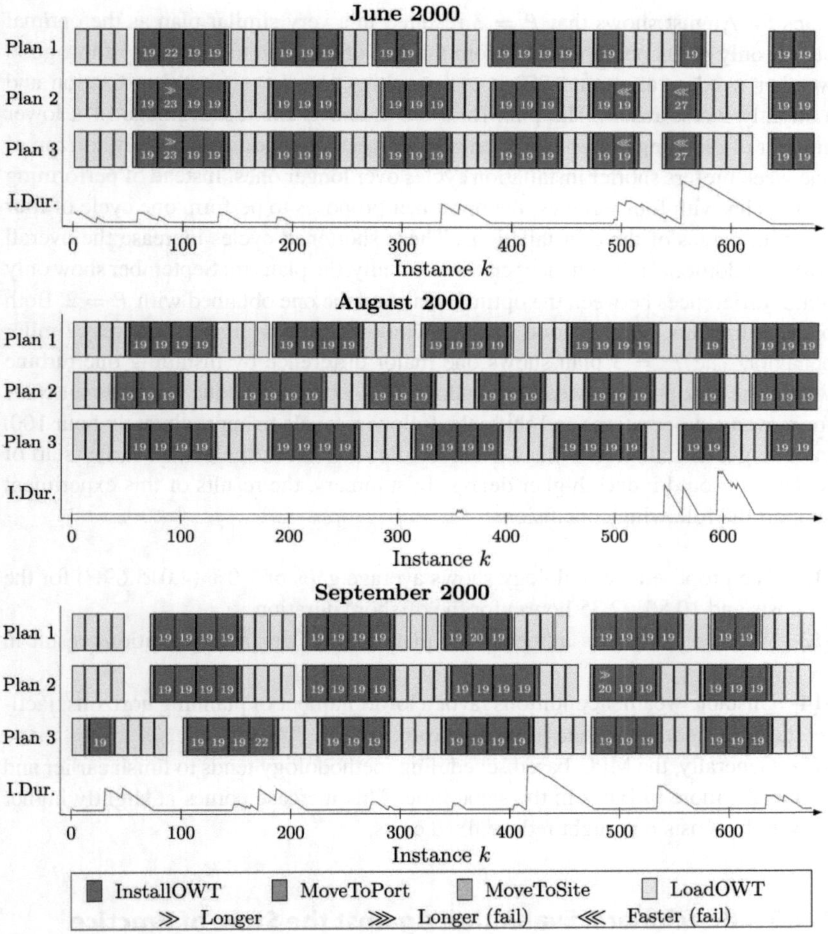

Fig. 8.7 Comparison with optimal plans

plans for August shows that $P = 3$ resulted in a very similar plan as the optimal one. It only shifts one operation from the end of the plan to utilize the short good weather window at around 550 hours, resulting in a lower vessel utilization and a slightly worse result. The plan for $P = 2$ shows the disadvantage of a lower number of planning horizons. Given perfect weather conditions, a short horizon of one week prefers shorter installation cycles over longer ones. Instead of performing four cycles with four turbines, the optimizer proposes to perform one cycle of four and four cycles of three installations. These shortened cycles increase the overall costs by additional movement operations. Finally, the plans for September show only minor differences between the optimal plan and the one obtained with $P = 2$. Both plans utilize this month's good weather windows quite well and show very similar behavior. The $P = 3$ plan shows one major difference by installing one turbine within the first possible weather window. This extra tour induces additional costs for movement operations and shifts the following cycle to approximately hour 100, including an installation with an extended duration. Nevertheless, an earlier start of this cycle would induce higher delays. In summary, the results of this experiment support the following conclusions:

-12- The proposed methodology shows average gaps of 7.0% (4,015.69€) for the costs and 10.8% (2.35 hours) for the offshore duration.

-13- Weather conditions influence the plan quality. Unstable conditions result in higher gaps.

-14- Unstable weather conditions favor a lower number of planning horizons, facilitating a frequent update of volatile forecasts.

-15- Generally, the MPC-based scheduling methodology tends to finish earlier and installs more turbines in the same time. This increase comes at slightly higher variable costs but might reduce fixed costs.

8.2.5 Benchmark: Evaluation Against the State of Practice

While the last section compared the MPC-based scheduling methodology's results with optimal plans and assessed potential losses, this experiment aims to benchmark the methodology to state of practice methods. Therefore, the experiment first uses the proposed methodology to generate an optimized plan (Opt) using the stagnation optimization mode. Afterward, it compares this plan with the two approaches implemented as part of the AnyLogic MAS described in Sect. 6.2. The first approach (Pri) serves as lower-bound for the benchmark. Every time that a vessel finishes an operation, it obtains a list of possible new actions. It then performs the one with the highest priority, i.e., loading or installing before movement operations. This strategy does not include weather forecasts and, thus, is expected to deliver the worst results. The second strategy (Cyc) uses forecasts to select the next installation cycle. Therefore, each agent uses the described forecasting mechanisms to construct possible cycles, loading and installing between one and four (limited by the vessel's capacity) turbines. It then selects the cycle with the lowest expected delays and executes the respective plan. This strategy aims to mimic current practices, where port operatives and the vessels' captains decide more or less ad-hoc, using forecasts and their own experience.

Tables 8.8 and 8.9 show the detailed results of the experiment. This experiment includes two variants of the base-scenario, simulating each variant for all three starting months. First, Table 8.8 shows the results for a fully stocked base port and one to three installation vessels. Second, Table 8.9 uses the default settings for the base port's inventory but only uses one or two installation vessels. Accordingly, the tables denote the different settings by the variant (Inf / Def for *infinite inventory* or *default scenario*), the starting month, and the number of vessels (V). The tables list the results for all three approaches for each setting. The final three columns show the overall project duration to install all 50 OWT, the sum of offshore times as another indicator for expensive operation delays, and the total costs induced by the performed operations. Moreover, the tables show the difference between the approaches and the MPC-based scheduling approach in percent. Thereby, a positive number indicates that the approach performed worse than the proposed methodology, while negative numbers indicate better performance.

Table 8.8 shows the results for unlimited inventory levels at the base port for one to three vessels and each month. The results show that the proposed methodology results in slightly shorter plans. On average, its plans shorten the project duration by 38 hours (2.77%) compared to the priority-based decision strategy and by 84 hours (5.54%) compared to the forecasting-based strategy. This reduction spans a range from −16 to 116 hours (priority) and 51 to 131 hours (forecasting). The table shows two instances where the priority-based strategy results in slightly shorter plans. Nevertheless, these come at a steep increase in the project's variable costs of 20.6% and 34.0%. When considering the average cost reduction, the table shows an average decrease of 18.79% compared to the priority-based strategy and 2.25% for the forecasting-based, state of practice strategy. Thereby, the table shows two instances in which the forecasting-based strategy resulted in lower costs of 1.05% and 0.15%, which, in turn, resulted in extended project durations.

Table 8.9 shows the results for the more realistic default scenario with limited inventory and capacity. The results show that, on average, the proposed methodology results in slightly longer project durations, ranging from differences between −198

Table 8.8 Benchmark with state of practice approaches—fully stocked base port

Setting	Meth.	**Proj. Duration [h]**		**Time Offshore [h]**		**Cost [€]**	
Inf-June-V1	Opt	1,780		1,108		2,793,600	
Inf-June-V1	Pri	1,893	(6.35%)	1,280	(15.52%)	3,084,000	(10.40%)
Inf-June-V1	Cyc	1,831	(2.87%)	1,089	(−1.71%)	2,764,200	(−1.05%)
Inf-June-V2	Opt	959		1,103		2,801,400	
Inf-June-V2	Pri	1,049	(9.38%)	1,341	(21.58%)	3,366,600	(20.18%)
Inf-June-V2	Cyc	1,062	(10.74%)	1,150	(4.26%)	2,898,000	(3.45%)
Inf-June-V3	Opt	764		1,122		2,845,200	
Inf-June-V3	Pri	748	(−2.09%)	1,318	(17.47%)	3,430,800	(20.58%)
Inf-June-V3	Cyc	846	(10.73%)	1,155	(2.94%)	2,892,600	(1.67%)
Inf-Aug.-V1	Opt	1,791		1,065		2,716,200	
Inf-Aug.-V1	Pri	1,907	(6.48%)	1,294	(21.50%)	3,109,200	(14.47%)
Inf-Aug.-V1	Cyc	1,861	(3.91%)	1,105	(3.76%)	2,783,400	(2.47%)
Inf-Aug.-V2	Opt	903		1,058		2,702,400	
Inf-Aug.-V2	Pri	922	(2.10%)	1,124	(6.24%)	2,888,400	(6.88%)
Inf-Aug.-V2	Cyc	1,034	(14.51%)	1,108	(4.73%)	2,788,800	(3.20%)
Inf-Aug.-V3	Opt	690		1,089		2,773,800	
Inf-Aug.-V3	Pri	705	(2.17%)	1,067	(−2.02%)	3,186,600	(14.88%)
Inf-Aug.-V3	Cyc	741	(7.39%)	1,100	(1.01%)	2,769,600	(−0.15%)
Inf-Sep.-V1	Opt	2,623		1,365		3,275,400	
Inf-Sep.-V1	Pri	2,618	(−0.19%)	2,005	(46.89%)	4,389,000	(34.00%)
Inf-Sep.-V1	Cyc	2,720	(3.70%)	1,512	(10.77%)	3,554,400	(8.52%)
Inf-Sep.-V2	Opt	1,066		1,073		2,741,400	
Inf-Sep.-V2	Pri	1,069	(0.28%)	1,409	(31.31%)	3,455,400	(26.05%)
Inf-Sep.-V2	Cyc	1,117	(4.78%)	1,092	(1.77%)	2,774,400	(1.20%)
Inf-Sep.-V3	Opt	730		1,093		2,783,400	
Inf-Sep.-V3	Pri	733	(0.41%)	1,280	(17.11%)	3,387,600	(21.71%)
Inf-Sep.-V3	Cyc	848	(16.16%)	1,103	(0.91%)	2,808,600	(0.91%)

Table 8.9 Benchmark with state of practice approaches—default scenario

Setting	Meth.	Proj. Duration [h]		Time Offshore [h]		**Cost [€]**	
Def-June-V1	Opt	1,824		1,071		2,722,200	
Def-June-V1	Pri	1,893	(3.78%)	1,280	(19.51%)	3,084,000	(13.29%)
Def-June-V1	Cyc	1,831	(0.38%)	1,089	(1.68%)	2,764,200	(1.54%)
Def-Aug.-V1	Opt	1,810		1,058		2,698,800	
Def-Aug.-V1	Pri	1,907	(5.36%)	1,294	(22.31%)	3,109,200	(15.21%)
Def-Aug.-V1	Cyc	1,861	(2.82%)	1,105	(4.44%)	2,783,400	(3.13%)
Def-Sep.-V1	Opt	2,719		1,172		2,928,000	
Def-Sep.-V1	Pri	2,618	(−3.71%)	2,005	(71.08%)	4,389,000	(49.90%)
Def-Sep.-V1	Cyc	2,720	(0.04%)	1,512	(29.01%)	3,554,400	(21.39%)
Def-June-V2	Opt	1,385		1,085		2,763,000	
Def-June-V2	Pri	1,385	(0.00%)	1,085	(0.00%)	2,763,000	(0.00%)
Def-June-V2	Cyc	1,325	(−4.33%)	1,236	(13.92%)	3,057,600	(10.66 %)
Def-Aug.-V2	Opt	1,373		1,069		2,728,200	
Def-Aug.-V2	Pri	1,333	(−2.91%)	1,117	(4.49%)	3,021,000	(10.73%)
Def-Aug.-V2	Cyc	1,351	(−1.60%)	1,089	(1.87%)	2,764,200	(1.32%)
Def-Sep.-V2	Opt	1,820		1,176		2,937,600	
Def-Sep.-V2	Pri	1,622	(−10.88 %)	1,871	(59.10 %)	4,419,000	(50.43%)
Def-Sep.-V2	Cyc	1,617	(−11.15 %)	1,452	(23.47%)	3,456,000	(17.65%)

hours and 97 hours for the priority-based strategy and −203 to 51 hours for the forecasting-based strategy. On average, these two strategies achieve a reduction of 29 hours or 1.39% for the priority-based strategy and 38 hours or 2.31% for the forecasting-based approach. On the other hand, these scenarios show an even higher decrease in project costs. Thereby, the MPC-based scheduling approach achieves average savings of 23.26% (667,900.00€) considering the priority-based strategy and 9.28% (267,000.00€) for the forecasting-based strategy.

Figure 8.8 depicts the difference between each approach and the MPC-based scheduling methodology averaged by each inventory setting (Inf/Def) and in total across all scenarios. The figure shows the averages described above and shows the average across all simulation runs (*All*) and both strategies (*Both*). The results show that the proposed methodology achieves an average reduction of the duration of 4.06% and a cost reduction of 5.06% across all scenarios compared to the state of practice, forecasting-based strategy. Thereby, the approach achieves an even higher

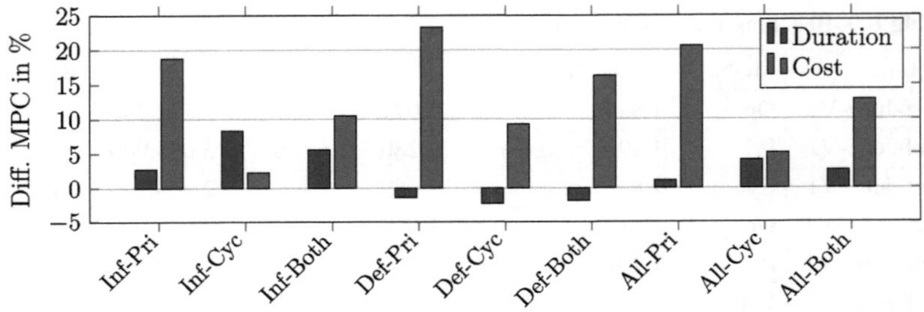

Fig. 8.8 Average differences by approach and inventory setting

cost reduction in the default scenario (9.28%) than in the infinite inventory setting. This behavior indicates that the proposed methodology shows advantages when the inventory is limited. Other experiments show a similar effect: The optimizer tends to extend the project duration if inventory levels drop very low by favoring cheaper over quicker installation cycles. Finally, the proposed methodology even achieves a cost reduction of 20.58% across all simulated instances compared to the assumed worst-case strategy (priority-based). These results allow drawing the following conclusions:

-16- The proposed methodology achieves very high cost reductions, averaging to 23.26% compared to the worst-case, priority-based decision strategy in the default scenario. This percentage amounts to an average reduction of 12,004.75€ per OWT.

-17- The proposed methodology achieves high cost reductions of an average of 9.28% compared to the state of practice, forecasting-based decision strategy in the realistic default scenario. This percentage amounts to an average reduction of 5,340.00€ per OWT.

-18- The results show higher cost reductions in the default scenario, indicating that the methodology handles limited inventory levels by favoring slower but less expensive installations.

8.2.6 Benchmark: Evaluation against the State of the Art

While the last experiment evaluates the proposed methodology regarding current state of practice approaches, this experiment provides a benchmark to current state of the art approaches. Unfortunately, the state of the art currently offers not methods that integrate current weather measurements and forecasts (c.f. Chap. 4). Nevertheless, the state of the art shows that scheduling applications generally focus on specific classes of methods, i.e., robust approaches or online optimization in terms of short-term control. In general, robust methods impose worst-case assumptions to reduce the impact of disruptions, while online control methods aim to find the next operation that should be conducted. In contrast to the proposed methodology, online control only proposes a single operation instead of a plan containing several actions. Considering this thesis's topic, a robust approach would integrate weather forecasts by assuming that the weather tends to the worst possible case given by the forecasts. In contrast, an online control approach mitigates the influence of forecast uncertainties by repeating the optimization after each operation, obtaining new forecasts for each iteration. This experiment modifies the proposed methodology to follow these two basic approaches, as literature proposes no methods for a direct benchmark:

Online Control: Online control only aims to propose a single action with a high frequency. The experiment directly applies the proposed MPC-based methodology as presented but modifies the sampling step width to a single hour ($T = 1$) and uses 168 prediction periods ($P = 168$). Consequently, this variant either proposes to start an operation or to wait for one hour. If it starts an operation, the next iteration starts after finishing the proposed one. The number of prediction periods follows previous results that this methodology requires at least a prediction horizon of one week ($N \geq 168$).

Robust Optimization: This variant of the proposed methodology modifies operation durations' estimation to follow a robust approach. While the original methodology tries to predict the duration as good as possible, the robust variant tries to ensure that it does not underestimate the duration. Therefore, this variant uses the sliding windows duration estimator, as described in Sect. 5.3.3.1, and selects a high value for the minimum required probability for each window $\omega = 0.9$. Consequently, it only accepts a window when the joint probability for viable weather conditions across all hours within this window matches or exceeds 90%. Moreover, this variant does not abort the plan execution if an operation finishes faster in the simulation than planned, as such situations are part of the general robust assumption.

The experiment compares the default scenario for the three starting months, June, August, and September, for all three variants. Table 8.10 shows the number of plans required for the scenarios, the projects' overall duration and costs, and the number of duration estimation errors.

Table 8.10 Comparison with state of the art approaches

Approach	Month	Plans	Faster	Slower	Too Slow	Duration	Cost [€]
Online	6	246	2	0	0	1793	2,706,000
Default	6	22	4	2	0	1824	2,722,200
Robust	6	24	2	0	0	2005	2,719,200
Online	8	273	0	0	0	1810	2,681,400
Default	8	20	0	0	0	1810	2,698,800
Robust	8	21	1	0	0	1875	2,743,200
Online	9	1154	0	0	0	2738	2,792,400
Default	9	33	2	1	1	2719	2,928,000
Robust	9	34	4	0	1	2825	2,998,200

The table clearly shows the difference between the three variants regarding the number of plans and wrongly estimated operation durations. While the default setting requires 22, 20, and 33 plans to finish the installation project, the online control variant requires 246, 273, and 1154 plans. These high numbers mainly result from bad weather periods, where this variant only waits for a single hour before calculating a new plan. In addition, the results show that only two of the planned operations turned out to require less time than expected. This result shows that this variant mitigates the influence of forecasting uncertainty to a high degree, but not completely. While such operations would trigger a recourse action in the default approach, the online control variant just continuous with the next iteration right after the operation finishes. The default variant shows a total of seven recourse actions: six operations finished faster than expected, and a single one took longer. Nevertheless, the online control variant offers no plans for project managers or operatives that they can rely on. While this variant shows the highest logistical performance, it reduces the predictability of operations. It requires operatives at vessels and at the port to react very quickly to new instructions on an hourly basis.

The robust variant only shows a slight increase in required plans to 24, 21, and 34 plans. This increase results from the extended project duration of, on average, 117.3 hours. In contrast, this variant only induces a single recourse action as one operation finished slower than expected, resulting in missing the next scheduled operation. Additionally, the results show that a total of seven operations finished faster than expected. Nevertheless, this variant considers these operations successfully conducted. Such overestimations constitute the expected behavior of robust assumptions. On the one hand, these results show an increased plan reliability of the robust assumption resulting from fewer plan disruptions. On the other hand, the results show a decreased logistics performance regarding the project's costs and duration.

Figure 8.9 depicts the results for the project duration and costs graphically for easier comparison. When comparing the project duration, the results show an increase in the robust approach, which averages 5.80% of the proposed methodology's results across all three months. In contrast, the online control approach shows very similar project durations, achieving an average decrease of 0.33%. Regarding the costs, the online control achieves an average cost reduction of 1.96% and shows a reduction in all three scenarios. The robust approach results in a slight increase in costs, averaging 1.31% across the three months. This increase results from the limited time windows available to the optimizer. The robust assumption introduces more time windows, where the duration estimator assumes longer installation times. Consequently, the optimizer chooses to avoid these windows, resulting in shorter installation cycles, i.e., in more cycles where the vessel only installs two or three turbines instead of four. These results allow drawing the following conclusions:

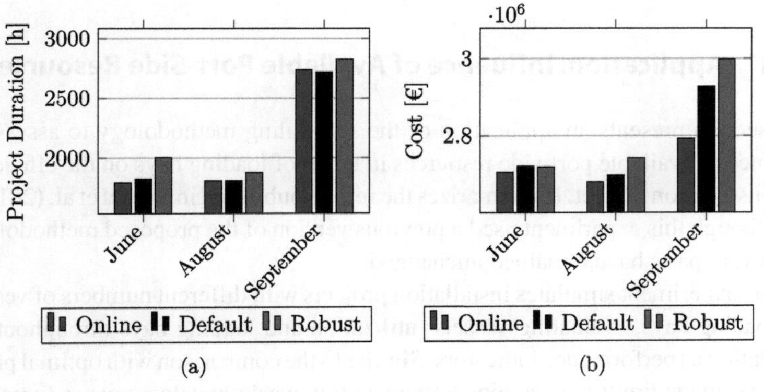

Fig. 8.9 Results of comparing with state of the art approaches

-19- Online control achieves better plans, leading to an average reduction of the costs (1.96%) and project durations (0.33%). In contrast, it does not offer any plans and, thus, nothing for planners and operatives to rely on.

-20- A robust assumption results in less efficient plans, leading to an average increase in the costs (1.31%) and project durations (5.80%). In contrast, it increases planning reliability. It only induced a single recourse action across all three scenarios compared to seven for the default variant.

8.3 Applications for Higher-Level Planning Tasks

This section presents three different applications of the proposed DSS, its underlying methodology, and components. While this thesis primarily aims to provide decision support on the operative level, the included methods and models also support planning tasks on higher levels. Therefore, this chapter applies various combinations of these methods for different planning tasks. First, it presents an experiment that mainly relies on the underlying MILP model to assess the influence of limited loading bays on the installation efficiency in Sect. 8.3.1. Second, it presents an application to optimize the required base port capacity in Sect. 8.3.2. Finally, it presents an evaluation of viable resupply cycles as input for the general optimization model and the capacity optimization in Sect. 8.3.3. This experiment includes all the models and methods presented in Chap. 5.

8.3.1 Application: Influence of Available Port-Side Resources

This section presents an application of the scheduling methodology to assess the influence of available port-side resources in terms of loading bays on the efficiency of an installation project. It summarizes the results published in Rippel et al. (2019b). Even though this experiment used a previous version of the proposed methodology, its relevant parts have remained unchanged.

This experiment simulates installation projects with different numbers of vessels and loading bays, evaluating vessels' utilization and loading bays throughout the installation as performance indicators. Similar to the comparison with optimal plans, this experiment limits the planning horizon to four weeks and does not use forecasts, i.e., the optimizer uses available historical weather recordings for a single period of 672 hours. This experiment uses the months August as a good-weather month, June with moderate conditions, and October 2000 to have as bad as possible, but still viable, weather conditions. This experiment uses October instead of September to have even worse conditions. The limited planning duration allows for this exchange. Finally, this experiment assumes a fully stocked base port to support a high number of installation vessels. This experiment simulates different ratios between vessels and loading bays to estimate the influence of these ratios. Moreover, the experiment includes several experiments with equal ratios to ensure that equal ratios also give comparable results.

Table 8.11 summarizes the experiment's settings and results. The table shows that similar ratios result in the same results. Additionally, Fig. 8.10 depicts the results graphically. The results show a saturation of the loading bay utilization, starting at a ratio of approximately three vessels per bay (0.33), indicating that additional vessels do not increase the bay utilization. This saturation results from the availability of viable weather windows within the scenario. The optimizer only schedules loading operations if it can also plan installations. The fact that the utilization caps at different levels for the three months, with higher values for better weather months, supports this assumption. Similarly, the results show a drastic decrease in vessel efficiency in terms of installed turbines per vessel from this point on. In terms of efficiency, all three scenarios maintain comparably high efficiency for ratios larger or equal to 0.5, i.e., a maximum of two vessels per bay. In conclusion, this experiment indicates that the application of two to three installation vessels per bay constitutes the maximum viable option. In particular, the strong decrease in vessel efficiency beyond this point will, in most situations, not warrant the additional costs for more vessels.

Table 8.11 Settings and results of the experiment to assess the influence of limited loading bays—adapted from Rippel et al. (2019b)

Ratio	Bays	Ves.	June Util.	June $\frac{OWT}{V}$	Aug. Util.	Aug. $\frac{OWT}{V}$	Oct. Util.	Oct. $\frac{OWT}{V}$	Mean Util.	Mean $\frac{OWT}{V}$
0.10	1	10	0.72	3.30	0.76	3.50	0.60	2.60	0.69	3.13
0.20	1	5	0.72	6.60	0.76	7.00	0.60	5.20	0.69	6.27
0.25	1	4	0.72	8.25	0.76	8.75	0.60	6.50	0.69	7.83
0.33	1	3	0.72	11.00	0.76	11.67	0.60	8.67	0.69	10.44
0.50	1	2	0.52	12.00	0.61	14.00	0.46	10.00	0.53	12.00
0.50	2	4	0.52	12.00	0.61	14.00	0.46	10.00	0.53	12.00
0.75	3	4	0.37	12.75	0.41	14.00	0.32	10.25	0.37	12.33
1.00	1	1	0.31	14.00	0.31	14.00	0.28	12.00	0.30	13.33
1.00	2	2	0.31	14.00	0.31	14.00	0.28	12.00	0.30	13.33
1.00	3	3	0.31	14.00	0.31	14.00	0.28	12.00	0.30	13.33
1.00	4	4	0.31	14.00	0.31	14.00	0.28	12.00	0.30	13.33

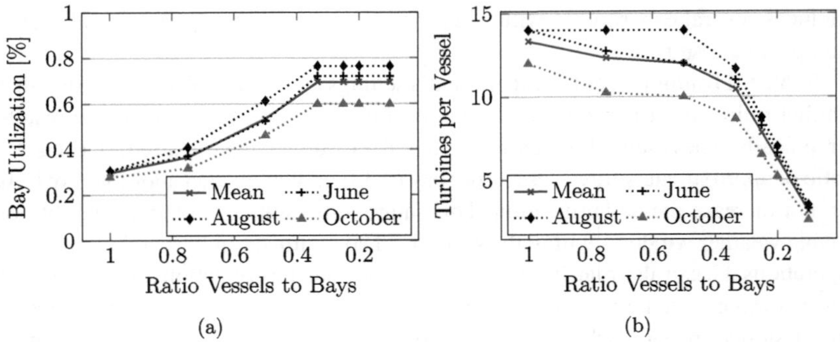

Fig. 8.10 Influence of the ratio between vessels and available loading bays—adapted from Rippel et al. (2019b)

This experiment shows an application of the proposed scheduling methodology and its components to assess the influence of loading bay availabilities on the installation process. While the experiment does not include forecasts and, thus, does not apply the MPC scheme, it uses various other components, e.g., the duration estimation or the underlying MILP model. These results show that the MPC-based scheduling methodology provides a viable foundation for extending the DSS's capabilities for offline optimization tasks. In general, this experiment allows drawing the following conclusions:

-21- The proposed methodology allows running various experiments considering installation projects' general characteristics by selecting the methodology's appropriate components. On strategic or tactical levels, the selected methods do not need to include the general MPC scheme as these levels cannot rely on current forecasts.

-22- The experiments show that two vessels per available bay constitute the best compromise, while three vessels still provide viable results.

8.3.2 Application: Optimization of the Base-Port's Capacity

This section summarizes an application of the scheduling methodology to determine optimal capacities for the base port, as published in Rippel et al. (2020a). The experiment uses the formulation presented in Sect. 5.4.2 and compares the presented approaches to estimate the output stream using historical data, i.e., the expected number of turbines built in a given time frame. Therefore, one approach applies the presented scheduling methodology to handle hourly aggregate data of N^{years} many historical years and the involved variance (scheduling). The other approach uses a statistical analysis of the data by constructing mean values over the required windows given by N^{years} and the duration of the given resupply cycle RES^{frq} (aggregation). This experiment serves two purposes. First, it serves as proof of concept that the proposed methodology can support higher-level planning tasks and works sufficiently well with aggregate historical data. Second, it benchmarks the proposed scheduling methodology with a statistics-based approach.

The experiment uses the geographical layout and the resupply cycle described in the base scenario. This scenario follows a use-case described by Beinke et al. (2017), which determines the relevant times for the resupply cycle RES^{frq} and the traveling times between the base port and the installation site. The experiment evaluates the setting given in Table 8.12. It evaluates both approaches for April, June, and August for five different years. It selects these months to increase the probability of having different weather compositions across these years. Generally, August tends to provide quite good conditions, while April generally tends towards relatively unstable conditions. June tends to provide moderate conditions in general. Finally, the experiment instructs the optimization to maintain three levels of minimum inventory to assess this parameter's influence and the algorithm's efficiency in maintaining these levels.

Table 8.12 Settings for the capacity optimization experiment (Rippel et al., 2020a)

Parameter	Notation	Domain
Year to evaluate		1980, 1985, 1990, 1995, 2000
Month to evaluate		April, June, August
Turbines to install	N^{owt}	50
Minimum Inv. Level	INV^{-}	$\{1, 2, 5\}$
Past Years to regard	N^{years}	20
Window Size (hours)	RES^{frq}	312
Restock amount	RES^{amt}	$in_l = 8, \forall l \in N^{rc}$

The experiment first applies both approaches to estimate the expected output flow at the base port. Therefore, it uses historical weather records, e.g., if planning for the year 2000, this step applies data for 1979–1999. Afterward, it obtains optimized capacity and initial inventory levels. Finally, it uses the MPC-based scheduling methodology to simulate the installation project. Therefore, it uses weather data from the current year, e.g., 2000, and generates relevant forecasts. The evaluation compares the results of both approaches in terms of their difference in plan length, capacity, and initial inventory level. As the experiment cannot obtain optimal values for these, it presents results relatively by subtracting the aggregation approach's results from the scheduling approach, denoted as (s-a). Thus, negative values show an advantage for the scheduling-based approach (s), while positive values denote an advantage for the aggregation-based approach (a). The evaluation focuses on the plan length to determine delays induced by empty storage, as the optimizer schedules loading and installation operations quite efficiently. It tries to circumvent long waiting times resulting from out-of-inventory events by delaying operations to achieve cost savings. Moreover, the experiment records the delay in shipments in hours, which results from overfull storage. Finally, the experiment records the highest and lowest storage levels throughout each simulation. It uses these values to determine adherence to the desired minimum inventory level. Moreover, it calculates the optimizer's capacity overestimation, i.e., the difference between the capacity and the used maximum storage.

Figure 8.11 shows the results of all 45 simulation runs, displaying the difference (s-a) in plan length (divided by 10), capacity, and overestimation. Accordingly, values below the zero line indicate a result favoring the scheduling based approach. Generally, these plots show a better performance of the scheduling-based approach, considering the number of instances and considering the magnitude of the results. Considering the project duration, the scheduling-based approach performed better in 14 instances with a sum of 372 hours, worse in 9 instances with a sum of 235 hours, and equally well in the remaining 22 instances. As noted, an increased plan length indicates an underestimation of the capacity. In sum, the scheduling-based approach proposed a lower capacity of 40 sets across all simulations. In detail, 69 sets less across 25 instances, compared to 29 sets across 16 instances where the aggregation-based approach performed better. Considering the overestimation, the scheduling-based approach performed better in 22 instances and only performed worse in three instances. Finally, the plots show a continuous decrease in the project duration/plan length with an increasing minimum inventory level. This decrease supports the assumption that this value provides a viable indicator for a capacity underestimation.

Table 8.13 summarizes the results as average values across all instances with the same minimal inventory INV^-. The first compartment shows the average values for the difference in plan length, capacity, and inventory level to enable a direct comparison between both approaches. These results show the scheduling-based approach, on average, proposes approximately one less capacity but, in turn, proposes a slightly

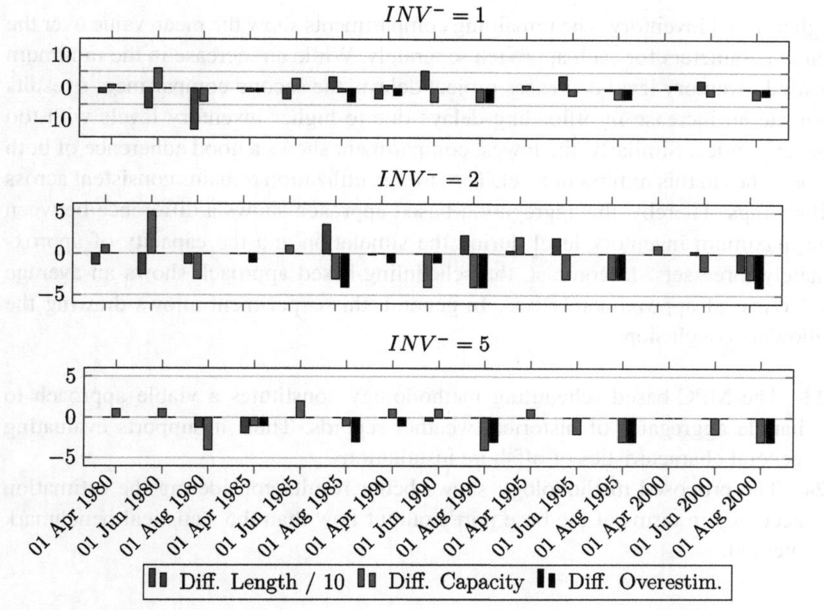

Fig. 8.11 Results of the capacity optimization by minimum inventory—derived from Rippel et al. (2020a) © 2020 SCS

Table 8.13 Results of the capacity optimization as mean values for different settings of the minimum inventory level Rippel et al. (2020a)

Parameter	Units	Approach	INV^-		
			1	2	5
Diff. Plan Length	[Hours]	(s-a)	−7.53	0.73	−2.33
Diff $Y^{capacity}$	[Sets]	(s-a)	−0.80	−1.00	−0.87
Diff $Y^{initial}$	[Sets]	(s-a)	0.27	0.27	0.27
Sum Delayed Restocks	[Hours]	(a)	20.33	4.20	40.73
Sum Delayed Restocks	[Hours]	(s)	0.00	31.73	0.00
$Y^{capacity}$—Sim. Max Inventory	[Sets]	(a)	3.33	3.40	3.47
$Y^{capacity}$—Sim. Max Inventory	[Sets]	(s)	2.33	2.20	2.53
Sim. Minimum Inventory	[Sets]	(a)	0.80	1.73	4.47
Sim. Minimum Inventory	[Sets]	(s)	0.80	2.20	4.53

higher initial inventory. The remaining compartments show the mean value over the stated parameters for each approach separately. While an increase in the minimum desired inventory level decreases project delays, the second compartment's results indicate an increase in offloading delays due to higher inventory levels with too low capacities. Similarly, the lowest compartment shows a good adherence of both approaches to this minimum level. The storage utilization remains consistent across all settings. Thereby, the aggregation-based approach shows a difference between the maximum inventory level during the simulation and the capacity of approximately three sets. In contrast, the scheduling-based approach shows an average difference of approximately two. In general, this experiment allows drawing the following conclusions:

-23- The MPC-based scheduling methodology constitutes a viable approach to handle aggregates of historical weather records. Thus, it supports evaluating general characteristics of offshore installations.

-24- The proposed methodology shows better results considering the estimation accuracy in terms of the base port's output flow than the statistical benchmark method.

8.3.3 Application: Optimization of Resupply-Routes

This section presents an experiment's results to determine the optimal resupply cycle for the described base scenario. The results have been published before in Rippel et al. (2020b). This experiment ties all the presented components of this thesis together by first optimizing different resupply cycles for the base scenario. Afterward, it determines the optimal base port capacity and initial inventory level for each possible cycle using the scheduling-based estimator. Thus, it determines options for viable input streams before estimating the corresponding output streams. Finally, the experiment simulates the optimized scenarios using the MPC-based scheduling methodology proposed in this thesis. As a general setup, the experiment tries to optimize a project starting in June 2000, as a month with moderate weather conditions. Thus, both preliminary optimizations use weather data for the years 1979 to 1999. The final simulations use weather data for the year 2000 to evaluate the results. As before, this experiment uses the presented user interface to generate the input scenario. Consequently, it only requires a single new Matlab script to tie all the models together. This script executes the generated scenario code and runs the transport optimization for an increasing number of round-trips N^{rt}. For each of these results, the script updates the state model before progressing to the capacity optimization. Finally, it updates the state model again and executes the simulation. This experiment serves several purposes. First, it evaluates the general feasibility of the proposed transport optimization model by applying it to the base-scenario for different values of N^{rt}. This part of the experiment allows drawing general conclusions about this value's influence on resupply cycles' efficiency. It offers an option to validate the model against the manually optimized resupply cycle described in Beinke et al. (2017). Second, this experiment constitutes another proof of concept that the proposed methodology provides a viable alternative to support planning tasks on higher levels. Finally, it demonstrates the proposed methodology's extensibility as a DSS by adding and combining several models for different tasks into a single experiment.

In summary, the transport optimization model proposed in this thesis solves a variant of a transshipment problem. These problem classes generally try to find a sequence of optimal cargo compositions (assignment) and routes to satisfy all orders given a network of sinks and sources. The model proposed in this thesis exploits the small size of the supply network to separate both problems. Therefore, it first solves a standard TSP for each possible combination of visited ports to obtain the round trips' duration. It then uses these durations to solve a custom formulation of a multi-periodic knapsack problem to obtain optimal cargo assignments. This formulation includes the duration of the assignment-dependent routes into its cost function and regards limitations concerning the HLV' payload and deck area capacities. The formulation tries to obtain as many complete sets of components using as little time as possible for a given number of allowed round-trips (periods).

8.3.3.1 Characteristics of Resupply Cycles and Validation

The first part of this experiment aims to assess the resupply cycles' general characteristics, considering the allowed number of round trips. Furthermore, this part includes a validation of the model. Therefore, this part compares the optimization results with the manually optimized resupply cycle described in Beinke et al. (2017). This reference cycle consists of four round-trips, fetching eight sets of components with a total duration of 310 hours.

While most of the parameters of the base-scenario result from this reference scenario, the authors do not provide all necessary values directly, as described in Sect. 7.1. In particular, the authors do not state information about the HLV's capacities and the components' dimensions but only provide information on how many components of each type the HLV can load simultaneously. The authors state that the HLV can transport eight towers in a single trip but only four nacelles. When calculating these limits from the averaged values given in the base-scenario, this thesis concludes the opposite: The HLV can transport eight nacelles but only four towers on a single trip. This discrepancy results in a slight difference between the optimizer's result and the reference cycle. While the reference cycle takes 310 hours, the optimizer's result takes 312 hours. This difference results from visiting a different port twice, as the scenario assumes towers to be manufactured in Cuxhaven while Nacelles originate from Bremerhaven. Adjusting for this difference, i.e., exchanging the production locations, results in a similar cycle duration of 310.6 hours (optimizer) compared to the 310 hours given in the reference cycle. Similarly, both cycles include a similar composition of cargo. In conclusion, the validation shows equal results between the proposed optimization model and the manually optimized cycle in Beinke et al. (2017).

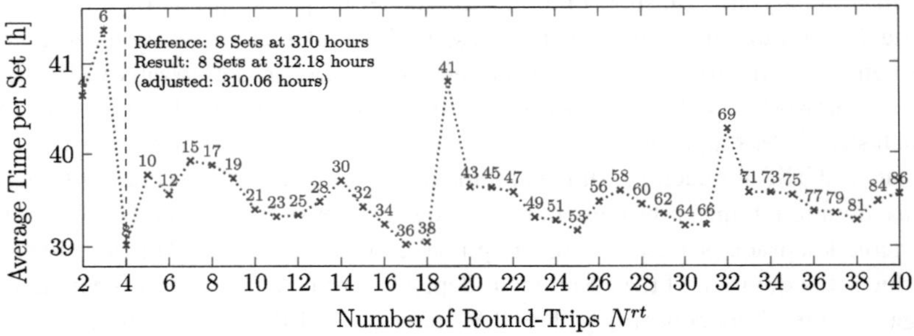

Fig. 8.12 Average time per delivered set concerning the number of allowed round-trips. Numbers at marks denote the total number of sets per cycle—derived from Rippel et al. (2020b) © 2020 IEEE

Figure 8.12 shows the average time per delivered component set for resupply cycles, including two to forty round-trips. The number at each mark shows the total amount of sets delivered in this cycle. The figure uses the average times per set, as these show more distinctive differences than the total time per cycle due to the smaller numbers. Moreover, these values facilitate a direct comparison of different cycles in terms of their overall efficiency. The figure omits $N^{rt} = 1$, as this setting induced a very high average time per set of more than 46 hours.

The results show a slight but continuous trend towards longer average times with an increasing number of allowed round-trips. Moreover, the figure shows a repeating pattern, starting at $N^{rt} = 3$, 19, and 32. Except for the first segment, each segment starts with a very high value before slowly dropping into a local minimum before the average time increases and decreases again into a second local minimum. The segment's internal differences decrease for each segment, indicating that the average duration might flatten for really high numbers of round-trips. Thereby, the first segment shows the highest variability and, in contrast to the following segments, shows an additional minimum at $N^{rt} = 4$. Moreover, this minimum also constitutes the global minimum, delivering eight sets in four round-trips with a total duration of 312.18 hours. These segments' existence and shape indicate that some settings allow the optimizer to determine more efficient resupply cycles than other settings. An analysis of the local minima shows that the involved cycles prefer transporting a single type of component, compared to visiting several ports on multiple trips. Nevertheless, this result highly depends on the selected supply network. On the one hand, these results show that a careful selection of the resupply cycles can increase their efficiency. Moreover, they show that a lower number of allowed round-trips results in generally more efficient resupply cycles. On the other hand, the results also show a comparably low variability in this duration of approximately two hours per set. In total, this results in a maximum delay of approximately 100 hours for all 50 sets in the considered scenario.

8.3.3.2 Evaluation of Resupply Cycles for the Base-Scenario

The first part of the experiment shows the resupply optimization's general characteristics that do not relate to specific weather conditions but only from the selected supply network and vessels. In contrast, the following evaluation tries to assess the impact of different resupply cycles on an installation project. Again, this experiment uses the base-scenario, as described earlier, but uses the transport optimization results as inputs for the capacity optimization and, again, for a simulation to evaluate the impact of different choices on the overall installation efficiency. Therefore, Table 8.14 shows the results for $N^{rt} = 1$ to $N^{rt} = 16$. Longer resupply cycles did not yield any feasible solutions during the capacity optimization. The required capacity and initial inventory level would need to exceed the turbines to install in these scenarios.

The table shows the results of each optimization or simulation stage in the different compartments. The first compartment again shows the transport optimization results in terms of the cycle duration, number of sets delivered, and the average time per cycle. The second compartment shows the capacity optimization's initial inventory level and capacity for each resupply cycle. Finally, the third compartment shows the simulation results in terms of the overall project duration, the minimum and maximum inventory levels during the simulation, and the number of hours that the storage was empty or that the HLV had to wait for unloading. The minimum and maximum storage values thereby ignore the first and last 200 hours to prevent them from dropping to 0 at the end of the simulation or returning the maximum inventory level at the beginning.

Table 8.14 Simulation results for the transport, capacity, and schedule optimization for $N^{rt} \in [1..16]$ (Rippel et al., 2020b)

	Transport			Capacity		Simulation				
Num. Trips	Cycle Dur. [h]	Num. Sets	Dur. / Set [h]	Init. Storage	Capacity	Proj. Duration [h]	Storage Min	Storage Max	Empty Storage [h]	Delayed Cycles [h]
1	92.1	2	46.0	27	27	1,792	11	26	0	0
2	162.6	4	40.7	18	18	1,792	6	17	0	0
3	248.2	6	41.4	17	17	1,792	2	16	0	0
4	312.2	8	39.0	17	17	1,792	1	16	0	0
5	397.7	10	39.8	15	15	1,828	0	14	150	0
6	474.8	12	39.6	18	18	1,792	0	17	8	0
7	598.9	15	39.9	29	29	1,792	7	28	0	0
8	677.9	17	39.9	21	21	1,823	0	20	135	0
9	755.0	19	39.7	27	27	1,792	2	26	0	0
10	827.5	21	39.4	33	34	1,792	6	32	0	0
11	904.5	23	39.3	40	40	1,792	11	39	0	0
12	983.6	25	39.3	46	48	1,792	17	45	0	0
13	1,105.7	28	39.5	34	35	1,792	2	33	0	0
14	1,191.2	30	39.7	38	40	1,792	3	37	0	0
15	1,261.8	32	39.4	43	45	1,792	5	42	0	0
16	1,334.3	34	39.2	48	49	1,792	8	47	0	0

On the one hand, the results show that none of the scenarios underestimates the required storage, as the HLV can unload its cargo in every case when it enters the port. On the other hand, three scenarios run out of components during the simulation. In two cases, this lack of components results in extended project durations. In the third case, this lack of components only extends over eight hours and does not influence the project duration. Overall, the results show a good usage of the selected capacity with a difference of one to three sets between the capacity and the maximum used storage. In contrast, the minimum used storage shows comparably high numbers for some runs, indicating that the optimizer chose the initial storage level to high in these settings. The experiment shows a decrease in the required capacity until $N^{rt} = 5$ followed by a trend towards increasing the capacity for higher numbers of round-trips. While $N^{rt} = 5$ shows the lowest capacity, it also induced a project delay, indicating that these values are too low. As a result, the scenarios with three or four round trips show the best overall results. In line with the reference scenario, this experiment concludes that selecting four round-trips for the resupply cycle constitutes the best option. This setting results in the highest cycle efficiency and shows the lowest discrepancy between the simulation's initial inventory level and capacity.

In conclusion, this experiment shows that the proposed methodology and the models proposed for the capacity and transport optimization obtain viable results using historical data. Therefore, the experiment used historical information and evaluated the different optimizations results using data for the "current" year. As a result, this experiment allows drawing the following conclusions:

-25- Considering the validation, the transport optimization model achieves the same cycle as the manually optimized resupply cycle presented by Beinke et al. (2017).

-26- The proposed MPC-based scheduling methodology serves as a viable foundation for decision support for higher-level planning tasks by combining it with other models. Thus, the methodology can support concrete installation projects even on higher levels.

-27- The results show that shorter resupply cycles show the best performance in terms of time per set and base port capacity. Nevertheless, too short cycles also increase the required capacity.

8.4 Summary

This chapter presents several experiments to evaluate the proposed DSS's components, the underlying methodology itself, and its viability for online optimization and higher-level planning tasks. The next subsections summarize the experiments' results and discuss their implications.

8.4.1 Summary of the Components' Results

Besides verifying the MAS, this chapter presents evaluations and comparisons of different options for the DSS's components. On the one hand, these experiments determined the most viable choice (**-1-**) or parameterization for this thesis (**-2-**, **-3-**). On the other hand, this chapter demonstrates the versatility of the proposed DSS's architecture. This thesis proposes alternatives for some of the components, where users could choose freely to apply one or the other approach. The results of this chapter show that these alternatives or alternative parameterizations show good results (**-1-**, **-4-**), whereby some choices result in better performances than others. Generally, result **-1-** shows that the use of interchangeable components, here in the form of duration estimators, provides the opportunity to enhance the overall DSS continuously. Similarly, the DSS supports exchanging other components, e.g., the function used to encode forecasting uncertainty, the duration estimator, the DSS's intervention policy, or even the fundamental scheduling formulation.

8.4.2 Summary of the Methodology's Results

This chapter provides several experiments to evaluate and benchmark the underlying scheduling methodology regarding its logistic and computational performance. The results show that the methodology provides a versatile framework for offshore wind installations' online planning under various circumstances. While the results show a required minimum prediction horizon of at least one week $N = 168$ (**-7-**), users can adapt their selection of the sampling step width T and the number of planning periods P to the current situation. On the one hand, shorter sampling step widths achieve better logistic performances, as they omit the influence of forecasting uncertainties (**-6-**). On the other hand, low sampling step widths result in shorter plans, which reduces the reliability of such installation projects, e.g., for personnel assignments. Consequently, users may need to select a viable trade-off between their desired plans' length, reliability, and performance. Therefore, the experiments show that using a sampling step width of $T = 84$ with $P = 2$ or $P = 3$ planning periods provide a good trade-off between logistic performance and computational requirements (**-8-**). Furthermore, the results imply that the number of planning periods partly depends on the expected weather conditions (**-13-**). While unstable weather conditions favor a lower number of planning periods to prevent highly unreliable forecasts (**-14-**), stable conditions favor higher numbers (**-5-**).

The results show that the methodology solves realistic scenarios, employing a single installation vessel and prediction horizons of approximately one week, optimally, even when using low optimization limits of 10 minutes (**-9-**). Nevertheless, the methodology cannot prove optimality for larger instances with more vessels or broader horizons within such low limits (**-10-**). On the one hand, these results show an increase in logistic performance with increasing limits. On the other hand, this increase only results in an average cost saving of an additional 3% if comparing the limits of one and thirty minutes for the default scenario using two installation vessels (**-11-**).

Besides evaluating different methodology settings, this chapter presents several benchmarks to assess its performance concerning other approaches. Even if the experiments resulted in different project costs due to the varying assumptions, the relative results provide a general frame to assess the proposed methodology's performance. Figure 8.13 shows a comparison of the worst-case (**-16-**), state of practice (**-17-**), state of the art (**-19-**, **-20-**), and optimality (**-12-**) benchmark experiments. The figure uses the proposed MPC-based scheduling methodology as a reference and the relative differences (percentage) from the experimental results. This figure's costs relate to the proposed methodology's average project costs in the state of practice benchmark experiment in Sect. 8.2.5.

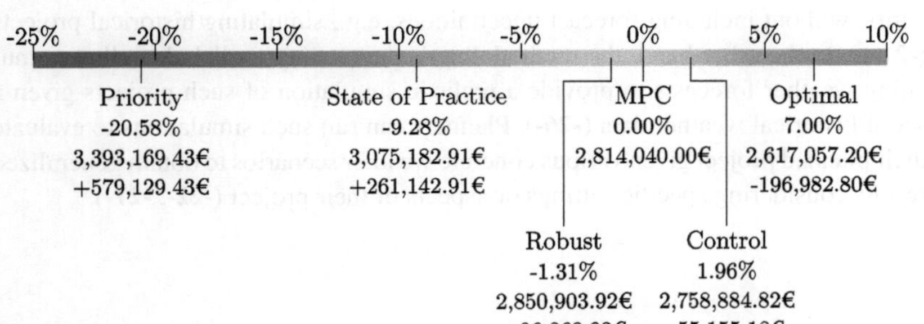

Fig. 8.13 Summary of benchmark results

The benchmarks show some specific characteristics of the proposed MPC-based scheduling methodology. On the one hand, it tries to finish the installation faster at slightly higher costs than the cost-optimal plans (**-15-**). On the other hand, the results show that the methodology works particularly well in settings with limited inventory levels. The applied MILP formulation, in combination with the MPC's predictive capabilities, allow the methodology to induce additional port-side waiting times in cases of a low inventory. While these waiting times increase the project duration, they reduce the overall costs by allowing more efficient but slower installation cycles (**-18-**). The proposed methodology's predictive-reactive assumption shows a trade-off between the high performance but non-existing plan reliability of the online control variant (**-19-**) and the lower logistic performance but higher reliability of the robust variant (**-20-**).

8.4.3 Summary of Offline Applications' Results

Finally, this chapter aims to demonstrate the viability of the proposed DSS and its underlying methodology for offline evaluations on the strategic or tactical level. Therefore, it presents several applications, or parts of it, to evaluate available port-side resources, transport, and capacity optimizations. The experiments show that the DSS offers a viable way to handle aggregate historical weather records and their involved uncertainties (**-23-**). Compared to a statistical benchmark method, the DSS results in higher quality solutions, e.g., during the capacity optimization (**-24-**), or similar results as manually optimized resupply cycles (**-25-**).

Generally, the experiments highlight that the underlying methodology can support a variety of tasks. On the one hand, it shows a high potential to optimize scenarios without including forecast uncertainties, e.g., simulating historical projects (**-21-**). On the other hand, the methodology's uncertainty model also allows emulating weather forecasts to provide a realistic simulation of such projects given a set of historical weather data (**-26-**). Planners can run such simulations to evaluate their planned project given various concrete weather scenarios to obtain generalized results considering specific settings or aspects of their project (**-22-**, **-27-**).

Conclusion 9

This chapter summarizes and discusses the proposed DSS and its underlying MPC-based methodology. Therefore, it reflects the last chapter's results regarding the imposed requirements in Sect. 9.1. Afterward, it presents a discussion of the research questions in Sect. 9.2 before it closes with a discussion of possible future work and open questions in Sect. 9.3.

9.1 Discussion of Requirements

This section aims to discuss the proposed methodology's characteristics regarding the functional requirements described in Sect. 2.4.2. Therefore, it discussed the previously presented results in terms of the stated requirements:

-R1- Solution quality, -R2- Real-time capability: The first requirement states that the DSS should aim at optimal results. It imposes the hypothesis that online scheduling might be limited by real-time constraints, while offline applications should result in mostly optimal results. The second requirement requires the DSS to generate plans in a reasonable time and a low amount of recourse actions to ensure reliable plans.

The proposed DSS uses a mathematical formulation combined with optimal branch-and-bound solution algorithms. It embeds this formulation into a receding horizon framework to reduce the problem size and allow for optimal solutions. The results show that this methodology solves instances with prediction horizons of one week within minutes, whether aiming at online or offline optimizations. Larger instances require a lot more time, even if the results show that short optimization times already result in close to optimal solutions. In terms of the real-time requirement, a regular offshore operation takes 2–3 hours. Usually, plan deviations

© The Author(s), under exclusive license to Springer Fachmedien Wiesbaden GmbH, part of Springer Nature 2026

D. Rippel, *Online Decision Support for Offshore Wind Farm Installations*,
https://doi.org/10.1007/978-3-658-49912-9_9

result from changing weather conditions, and thus, operatives should be able to anticipate such disruptions comparably early. Assuming that operatives observe a weather shift at the beginning of an operation, the extended operation times suffice to obtain a new follow-up plan. Regarding the optimality of solutions, the methodology obtains optimal plans for the default prediction horizon of one week. Nevertheless, the benchmark with optimized plans shows that the receding horizon can reduce the logistic performance compared to an optimal, non-forecast dependent plan. Thus, the methodology presents a trade-off between logistic performance (optimality) and the scheduling problem's solvability. Therefore, the resulting DSS allows modifying the sampling step width T and the number of planning periods P, adapting to available hardware, and the desired trade-off. Consequently, this thesis considers the real-time requirement as fully satisfied while considering the optimality requirement as mostly satisfied. On the one hand, the methodology obtains optimal schedules for realistic planning horizons. On the other hand, the receding horizon and the forecast uncertainty can still result in sub-optimal plans compared to plans, which do not suffer these restrictions.

-R5- Weather forecasts (online), -R6- Reactivity (online): These two requirements focus on capabilities necessary to achieve viable decision support during an installation project. On the one hand, the DSS needs to incorporate current weather forecasts to achieve reliable plans. On the other hand, it requires proper monitoring and intervention policies to react to changing conditions or wrongly estimated operation durations.

The DSS applies the MPC control scheme to achieve high reactivity while still providing reliable plans. It uses a novel approach to embed forecasting uncertainties into the forecasts. This approach converts discrete forecasts into probability intervals and applies one of two proposed duration estimators. These estimators try to provide a realistic estimation of an operation's duration and result in a predictive-reactive methodology. In conclusion, this approach differs from typical MPC applications in two ways: First, it provides a plan instead of a single control value (operation) as most approaches described in the state of the art do. Second, most of these approaches do not explicitly handle forecasting uncertainties and only rely on the receding horizon to omit them reactively. In conclusion, this thesis assumes both requirements to be fulfilled: First, by including dedicated methods to handle forecasts and their uncertainty. Second, by monitoring the plan execution and reacting in cases of plan failures. Thereby, the thesis assumes that more sophisticated recourse actions could enhance reactivity. Currently, the methodology simply starts a new iteration to obtain a new, optimized plan. Applying methods to "repair" the failed plan could enhance this process. While the methodology's default settings do not require such techniques, long planning horizons might benefit from such approaches as rescheduling quickly becomes time intensive with broader horizons.

-R3- Planning tasks with diff. horizons, -R7- Historical weather data (offline), -R8- Long-term schedule (offline): The first requirement states that the DSS should apply to higher-level planning tasks as offline optimization. In this context, the DSS needs to include approaches to aggregate and use historical weather records (**-R7-**) and the capability to obtain plans for complete projects (**-R8-**).

Section 8.3 shows several offline applications, while Sect. 8.1 shows an evaluation of those components that handle and aggregate historical weather records. These results show the scheduling methodology's versatility, both in providing inputs for other tasks (capacity optimization) or as a base-line methodology for various "what-if" analyses, e.g., to determine the influence of available loading bays. Thus, this thesis considers the first requirement fully satisfied. The results show that the proposed methodology constitutes a feasible foundation for integrating various online and offline optimizations into the DSS. The same holds for the third requirement. These applications show that the DSS can simulate complete projects using historical data. Finally, the experiments show two approaches to aggregate historical weather and use the resulting aggregates for such simulations. Both approaches, i.e., the analytical, windowed aggregation, and the hourly, scheduling-based approach, show promising results for the evaluated tasks. Nevertheless, this thesis assumes that this requirement is mostly satisfied. While the DSS shows good results with these basic methods, more advanced methods could increase accuracy. These methods could include machine learning models or other, more sophisticated prediction models.

-R4- Project focused interface: This requirement requested that project managers or operatives use the scheduling methodology as part of a DSS. This user group mainly requires the DSS to provide a project focused user interface that allows users to specify the scenario without in-depth knowledge about the underlying mathematical formulations or methods.

The DSS satisfies this requirement by using model transformations to detach the user inputs from the model formulation and code. While out-of-scope for this thesis, this detachment allows designing and implementing dedicated interfaces for each user group. Even if this thesis only proposes a basic interface, the underlying model transformation and meta-modeling approach satisfy the requirement for project-focused interfaces completely.

In conclusion, it can be stated that the proposed DSS and the underlying MPC-based scheduling methodology presented in this thesis satisfy all functional requirements to a high degree. Nevertheless, several aspects would benefit from enhanced methods, e.g., formulations that allow for larger prediction horizons, more focused recourse actions, or prediction-based methods to handle historical weather data. Table 9.1 summarizes the requirements and their degree of fulfillment, as assumed by this thesis.

Table 9.1 Requirements and degree of fulfillment

Requirement	Deg.	Requirement	Deg.
-R1- Solution quality	◕	**-R5-** Weather forecasts (online)	●
-R2- Real-time capability	●	**-R6-** Reactivity (online)	◕
-R3- Planning with diff. horizons	●	**-R7-** Hist. weather data (offline)	◕
-R4- Project focused interface	●	**-R8-** Long-term schedule (offline)	●

9.2 Discussion of Research Questions

While the last section focuses on discussing the proposed DSS in terms of the functional design requirements derived from the application domain, this section evaluates the research questions derived from the state of the art. In total, this thesis imposes four of these theoretical research questions:

RQ1 *Is it possible to apply Model Predictive Control to the online scheduling of offshore operations in a planning context (not just control)? Which methods and models does it require to act as DSS?*

The results presented in this thesis show that MPC provides a reliable foundation for planning tasks in dynamic environments. Most applications in the literature apply MPC to mitigate external influences and, thus, rely on a control approach to regularly revise their decisions. In contrast, this thesis proposes a methodology that includes these external dynamics, i.e., weather dynamics and uncertainties, into the MPC. This integration allows deriving plans instead of control decisions, covering several operations, and spanning more extended time frames.

The results also show that mainly two aspects limit this planning capability. First, the selected optimization model strongly limits the real-time constraint usually imposed on MPC applications. This thesis uses a MILP formulation, which provides viable results in short times. Nevertheless, an increasing prediction or planning horizon results in a massive increase in the required computational time. Second, planning in uncertain domains requires a precise estimation or model of the involved dynamics. The proposed scheduling methodology includes weather dynamics by estimating the duration of operations using provided forecasts. It converts these forecasts into probability intervals by encoding the a-priori known uncertainty into the forecast. As a result, forecasts with horizons of more than two weeks become increasingly uncertain and limits the possibility of planning further ahead. Consequently, longer horizons would require enhanced duration estimation techniques, e.g., by additionally incorporating historical weather.

RQ2 *Does a combination of receding horizon techniques, particularly Model Predictive Control and Mixed-Integer Linear Programming scheduling models, provide advantages over classical scheduling approaches, e.g., stochastic or robust optimizations?*

The results and benchmarks highlight a series of advantages over classical approaches. The state of the art shows that scheduling problems usually apply MILP formulations, as they include assignment problems. Thus, this research question focuses on the fundamental assumption of following a robust, stochastic, or, as in this thesis, a predictive-reactive approach. Stochastic approaches often follow a similar segmentation into smaller planning horizons as used in this thesis. Thereby, they solve several instances for each segment, representing different samples of the underlying stochastic variables. Robust approaches may or may not apply such a segmentation. However, the limited length of weather forecasts would probably imply such a segmentation for robust approaches in this context.

The results show that the proposed predictive-reactive approach achieves plans with a shorter project duration and lower costs than a robust optimization. In contrast, robust plans achieve higher plan reliability with slightly fewer recourse actions. Consequently, users need to decide which objective they rank higher, plan reliability, or cost-efficiency. As shown in the benchmark section, the proposed DSS and its underlying methodology can implement both approaches, robust or predictive-reactive, by selecting the corresponding duration estimator and configure them accordingly.

This thesis does not include a direct benchmark with stochastic approaches. Stochastic approaches rely on solving several instances of the scheduling problem for each time step. Thereby, the number of instances increases with an increasing prediction depth. For example, If each stage includes three samples (best-case, expected-case, worst-case), the second stage already needs to solve nine problems, while the number increases to 27 instances on the third stage. As scheduling problems at least belong to the class of NP-hard problems, solving such high numbers of instances would probably violate the DSS's real-time requirement. Nevertheless, this thesis assumes that the results might vary between the control-based (−1.9% costs) and the robust approach (+1.3% costs) given in Sect. 8.2.6. The exact results will depend on the segment's length, ranging from a single operation (control-based) to more extended plans, and the number and strategy of sampling.

In conclusion, this thesis shows that a predictive-reactive online-scheduling approach combining MPC and MILP shows several advantages over classic robust or stochastic settings. The proposed DSS constitutes a viable trade-off between the computational requirements, plan reliability, and cost-efficiency.

RQ3 *Is it possible to apply the proposed scheduling algorithm to historical weather records to achieve suitable approximations for long-term planning tasks? Which additional preprocessing does such an application require?*

The proposed scheduling methodology shows distinctive characteristics, rendering it a good option to work with historical data records. Generally, it replaces forecast data with aggregates of historical data. Therefore, methodology relies on averaged values as most of the other state of the art approaches do, but it can additionally incorporate the data's variance.

This thesis proposes two ways to aggregate such data by using windows of different widths. The used windows range from a single hour to several weeks. Even though both components only provide the average values for the experiments, they already achieve satisfactory offline planning results. Both experiments incorporate a large number of past years (20). Such large data would require a distinct approach to obtain sensible limits for the probability intervals from the data's variance. Preliminary experiments (not given in this thesis) show that incorporating the variance over these 20 years results in very similar probability intervals for all data points. Thus, a dedicated component that includes the variance could, e.g., weigh the data so that older data has a lower influence on the resulting intervals.

In conclusion, it can be stated that the proposed scheduling methodology is highly suited to work with historical data. At the same time, its modular design allows users to apply different approaches to aggregate this data. Nevertheless, the development and evaluation of more advanced aggregation components remain an open question for future work.

RQ4 *To which extend is it possible to use model transformation and code generation to provide project managers with a detailed and intuitive interface for the mathematical scheduling algorithm? Can other planning tasks be integrated into such an interface?*

This thesis proposes a model transformation framework to separate the algorithms from the user inputs. This way, users can define their scenario without altering internal data structures and without an in-depth knowledge of the underlying methods. The thesis demonstrates this process by transforming the same scenario definition for two very different target platforms, i.e., a MAS implemented in AnyLogic and the proposed scheduling methodology that uses a MILP. Consequently, it answers this research question with a definite yes.

Nevertheless, if users want to change the methodology's default settings, they need advanced knowledge about the related parameters and the DSS's inner workings. Therefore, future work will need to define which users might work with the DSS, e.g., logistics and process experts, methods experts, or management personnel. Future work needs to design designated interfaces for these user groups to allow displaying and modifying relevant information. Moreover, the current implementation aims to define and transform the users' scenarios. Particularly management personnel, require additional information on the current and predicted state of the project, which the current meta-model does not cover.

In conclusion, this thesis assumes that model transformations provide a viable option to facilitate the usage of a DSS for different user groups. Moreover, it provides a versatile approach to integrate additional methods, e.g., simulations or other optimization models, to extend the DSS's capabilities.

9.3 Open Research Questions and Possible Future Work

This section summarizes open research questions or ideas for extensions of the proposed methodology. Thereby, most of these ideas result from the discussions of the design requirements and the research questions.

Validation with real-world forecasts This thesis uses real-world weather measurements to simulate the planned installation projects. Unfortunately, this dataset only comprises measurements but no forecasts. Consequently, future work will focus on obtaining a matching dataset containing both measurements and forecasts. The results presented in this thesis use forecasts generated using the actual measurements, probably resulting in higher accuracy than real forecasts. Nevertheless, the DSS explicitly encodes uncertainties into the forecast values by generating increasing probability intervals. On the one hand, these uncertainties follow realistic assumptions. On the other hand, this encoding masks the actual measurements or forecast values from the underlying methodology. Consequently, this thesis imposes the hypothesis that this process results in comparable intervals for the currently simulated forecasts and actual real-world forecasts. Forecasting errors increase with increasing horizons. Similarly, the encoded uncertainty increases, which should result in comparable intervals.

Enhanced duration estimators: The currently proposed estimators focus solely on using forecasts to estimate the duration of offshore operations and either use a sliding window or a Markov-Chain-based approach. While both estimators show good results, enhanced versions could include Machine Learning to determine probable trends and incorporate these in the predictions. Moreover, enhanced estimators could include historical data to allow estimations beyond the forecasting horizon.

Using an evidence theoretic representation of weather conditions could enable a feasible approach to combine historical data, current measurements, and forecasts. The advantage of such a representation shows in its ability to represent uncertainty directly. For example, while the current estimators use a binary representation (weather is good enough or too bad), Dempster-Shafer Belief-Networks, and their derivatives use a different notion. Generally, these networks use a set of hypotheses (Frame of Discernment X). Afterward, they construct the powerset over X as hypothesis-space. In this example, the network could express the belief in the two hypotheses, that the weather will be good enough (A) or too bad (B), but additionally allow the notion of "I don't know" ($A \vee B$) as part of the power set. Using such a network allows creating a prior distribution using historical data, factoring in the observations (good/bad) over several past years. Therefore, they allow weighting historical data, e.g., by increasing the uncertainty (hypothesis $A \vee B$) with an increasing distance from the current year. Afterward, they allow updating the belief-values when adding new measurements and forecasts. Similar to the treatment of historical data, these networks could incorporate the uncertainty of forecasts directly by utilizing their third base-hypothesis. Future work should evaluate if such a representation proves beneficial. On the one hand, Dempster-Safer Networks tend to become large and update slowly if they include too many hypotheses. Particularly, if a single network covers the whole prediction horizon of, e.g., 168 hours, the network would consist of 2^{336} hypotheses. On the other hand, the dedicated expression of uncertainty and the ability to combine beliefs could increase prediction accuracy.

Enhanced aggregators for historical data: This thesis proposes two ways to aggregate historical data using windowed mean values. While these aggregators already provide sufficient accuracy for higher-level planning tasks, enhanced aggregators could result in higher precision. Therefore, the aggregators should incorporate the variance of historical data, as described in the discussion of the research questions in Sect. 9.2. Weighting this data according to its recency could result in more reliable probability intervals for the duration estimation. Other aggregators could apply methods that model the behavior of such data using its frequency or auto-correlation.

Precedence-based scheduling: Currently, the DSS applies a time-indexed MILP scheduling formulation. This thesis chooses such a formulation for easier integration with the MPC scheme, as it allows defining a time frame in which the optimizer schedules as many operations as possible. In contrast, precedence-based formulations use a time-continuous representation to schedule all operations. Therefore, these formulations require a sufficiently large time frame and need an estimation of operation durations for the whole project, which generally spans several months. Unfortunately, forecast uncertainties increase with an extended prediction horizon and render such estimations over several months infeasible. Nevertheless, such formulations provide increased scheduling performances, allowing to solve problem instances with broader horizons or with more vessels.

Future work should investigate if a precedence-based formulation could be adapted beneficially to this application. Suitable formulations categorize as "scheduling with time-dependent processing times". Alidaee and Womer (1999) provide a review of such formulations and highlight that these require a function for the processing times. The realization of such a function poses three major challenges: First, instead of a discrete look-up table, as applied by the current approach, the time-continuous nature of such formulations requires a time-continuous mapping of the processing times. Applying a piece-wise linear transformation to the table could achieve such a representation but might introduce inaccuracies. Second, such formulations represent operation start times as optimization variables instead of fixed indices. Therefore, these formulations need to represent the processing time function as constraints instead of an indexed, external parameter. Consequently, the methodology needs to generate constraints for each piece of that function automatically. Figure 9.1 shows such a piece-wise representation for the first week of June 2000. It shows each segment a solid line and indicates additional interpolations as a dotted line. This example shows that a single week with moderate conditions already consists of 15 segments plus additional segments to interpolate between these. Each segment requires at least three constraints to represent the logical implications that an operation uses this segment if started after the segments start and finishes before its end. Consequently, such a formulation would require a large number of constraints to represent the whole (possible) duration of such a project. Third, the methodology requires a customized duration estimator that handles the increasing uncertainty. As described before, forecasts only remain reliable for a short window. After that point, the estimator could resort to estimations using historical information to provide viable durations. On the one hand, such a procedure would render the estimations unreliable with an increasing horizon but would allow the formulation to schedule the entire project. On the other hand, the integration with the MPC scheme could render this unreliability unproblematic by applying short-term plans that still rely on the current forecasts.

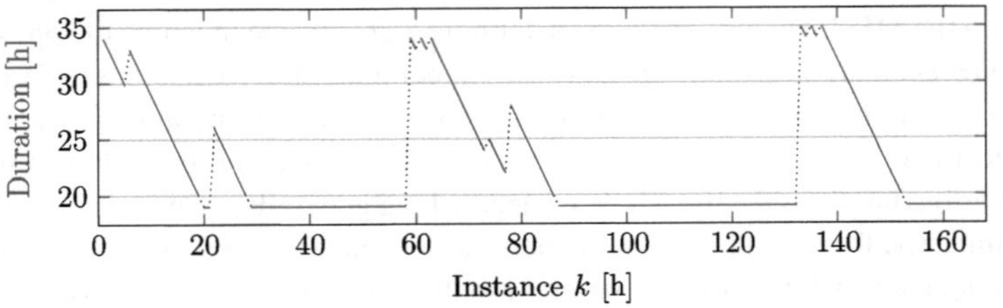

Fig. 9.1 Piece-wise linear segments for operation durations

In conclusion, this extension would require a reformulation of the scheduling model, a customized duration estimator, and a method to extract viable constraints from the duration estimator's results. As a research hypothesis, this extension could render broader prediction horizons feasible. In contrast, the increasing uncertainty induced by using historical data instead of forecasts, the extended prediction horizon, and the high increase in required constraints might render the performance improvements of such formulations irrelevant.

Enhanced Recourse Actions / Solution Algorithms: Currently, the proposed DSS uses an intervention policy that simply obtains a new plan if the current plan fails. While this policy suffices for the currently proposed prediction horizon and also shows viable results for larger problem instances, future work should investigate ways to include more distinctive recourse actions. Several rescheduling approaches exist in the literature, mainly relying on heuristic methods to solve a new plan. The current implementation uses CPLEX's precisely-solving branch-and-bound algorithm. Consequently, customized heuristics can achieve a large increase in efficiency. Nevertheless, considering the long operation durations involved, future work should examine the trade-off between an increased speed and the loss of solution quality. This thesis assumes that such approaches' main benefit would result from the ability to apply prediction horizons. Therefore, the accuracy of forecasts and the used duration estimators limit the length of sensible prediction horizons.

Interface design and stakeholder analysis: In its current state, the proposed DSS uses a simple, generated user interface that separates the scenario from the methods' settings. The current interface and data model comprise all planning and scheduling tasks covered in this thesis. Nevertheless, this thesis considers the design of these interfaces out-of-scope, and future work needs to focus on focal user groups. Therefore, it needs to establish, which user groups will likely use such a system, which information they need for their work, and how the information should be presented.

This thesis expects that, e.g., process experts, methods experts, or management personnel require a different set of information, represented in very different ways. Accordingly, future work will need to extend the underlying meta-model to include additional information relevant to management personnel. Moreover, future work will focus on transforming and managing the results of experiments or simulations back into the DSS's interface. Therefore, Text-To-Model transformations provide a suitable method to obtain these results. In its current state, the meta-model only holds a single state, i.e., the planned scenario or experiment to conduct. While it could easily update this state to the simulation or optimization state or final result, future work needs to establish additional ways to examine predicted states to allow manual monitoring of the project.

Personnel planning: The personnel planning and scheduling constitutes one major aspect in the construction of OWFs. This thesis considers this aspect out-of-scope to keep the tackled problem manageable. Nevertheless, future work will focus on this topic. On the one hand, construction operations require highly specialized offshore personnel that companies either employ on their own or charter from external companies. On the other hand, offshore personnel's schedules need to follow extensive legislative rules and restrictions, ranging from safety regulations over workforce regulations, e.g., pause- and working-times, to international regulations regarding visa and travel laws. Besides the need to maintain a sufficient workforce for scheduled operations, discussions with experts highlighted that assembly teams' experience highly influences the duration of operations. Consequently, a DSS needs to plan workforce availability while incorporating their experience into the scheduling itself.

In conclusion, this thesis presents a methodology for the online scheduling of offshore operations that explicitly considers forecast uncertainties. The proposed methodology shows very good results compared to the current state of practice approaches. This thesis's various experiments show the methodology's versatility and its applicability for planning tasks on the strategic, tactical, and operative levels. Nevertheless, the modular setup of this methodology provides various opportunities to extend or enhance the resulting DSS.

9.4 Transferability of the Proposed Methodology

The proposed DSS explicitly focuses on the installation of offshore wind farms, especially on their so-called top-structures. While it focuses on the requirements and restrictions of this domain, the underlying methodology uses a versatile combination of different components and shows high adaptivity. The methodology represents most of the domain's processes and their requirements using parameters exposed to the user. This includes the definition of basic and aggregate operations, their weather limitations, and durations. Consequently, users can easily adapt the DSS to different processes without the need to exchange models, methods, or other components or only with minor changes.

For example, apart from defining the respective processes, an extension to the first phase of constructing foundations requires only minimal modifications to the scheduling model. This phase uses two different vessels for the installation of foundations and cables. Consequently, the model needs additional constraints to assign specific sets of operations to specific vessel types. Nevertheless, this addition only requires constraints widely used in flexible job shop scheduling models. Inclusion of the commissioning phase requires minor modifications to the duration estimators. In contrast to the first two phases, the commissioning requires high wind speeds to test the turbines. Accordingly, the estimator needs to find suitable windows with high enough wind speeds.

When adapting the methodology to this use-case, the current formulation will determine optimal trade-offs between low transportation and waiting/processing times. This application might be particularly interesting when considering additional transportation modes (e.g., crew vessels or helicopters) with their respective weather restrictions and costs. This question often arises in the operations and maintenance phase of an OWF. In this phase, planners often face the problem of scheduling planned and unplanned maintenance activities, sometimes across several wind farms, using a fixed pool of vessels, helicopters, and engineers. At this, various maintenance tasks subject to different weather limits, requiring a bandwidth of weather conditions between calm (e.g., blade inspection and repairs) to windy (e.g., exchange and test of internal electronics). Consequently, the proposed methodology might provide a viable tool for this phase by incorporating current weather information and forecasts.

The proposed methodology should be applicable to other installation or maintenance activities in the offshore domain or generally subject to weather influences in a broader sense. The construction of offshore oil and gas rigs, pipelines, or underwater cables all subject to hardly predict weather conditions in one way or another. Similarly, several onshore construction projects exist that subject to weather limitations. Specifically, projects taking place in great heights, e.g., skyscrapers' construction, face similar problems. In a broader sense, the algorithms can be adapted to other weather conditions, e.g., frost or rain, which impede several construction processes.

Even though the proposed methodology could theoretically handle various problems that include time-dependent processing times under uncertainty, the current time-indexed optimization model limits the methodology's applicability to short- to mid-term planning tasks. Depending on the domain, the methodology might circumvent this restriction by selecting larger instances, e.g., days instead of hours, but this highly depends on the scheduled operations, dynamics, and uncertainties.

Bibliography

Abdolrazzagh-Nezhad, M. and Abdullah, S. (2017). Job shop scheduling: Classification, constraints and objective functions. *International Journal of Computer and Information Engineering*, 11(4):429–434.

Agadaga, G. O. and Akpan, N. P. (2017). Transshipment problem and its variants: A review. *Mathematical Theory and Modeling*, 7(5):19–32.

Ahn, D., Shin, S.-C., Kim, S.-Y., Kharoufi, H., and Kim, H.-C. (2017). Comparative evaluation of different offshore wind turbine installation vessels for korean west–south wind farm. *International Journal of Naval Architecture and Ocean Engineering*, 9(1):45–54.

Ait Alla, A., Oelker, S., Lewandowski, M., Freitag, M., and Thoben, K.-D. (2017). A study of new installation concepts of offshore wind farms by means of simulation model. In Chung, J. S., Triantafyllou, M. S., and Langen, I., editors, *Proceedings of the Twenty-seventh International Ocean and Polar Engineering Conference (ISOPE)*, pages 607–612. ISOPE:Cupertino.

Ait Alla, A., Quandt, M., and Lütjen, M. (2013). Simulation-based aggregate installation planning of offshore wind farms. *International Journal of Energy*, 7(2):23–30.

Alidaee, B. and Womer, N. (1999). Scheduling with time dependent processing times: Review and extensions. *The Journal of the Operational Research Society*, 50(7):711–720.

Arbeitsgruppe Erneuerbare Energien (2020). Time series for the development of renewable energy sources in germany—februrary 2020. Technical report. https://www.erneuerbare-energien.de/EE/Navigation/DE/Service/Erneuerbare_Energien_in_Zahlen/Zeitreihen/zeitreihen.html, last accessed 07.04.2020.

Arshad, M. and O'Kelly, B. C. (2013). Offshore wind-turbine structures: A review. *Proceedings of the Institution of Civil Engineers—Energy*, 166(4):139–152.

Assimizele, B., Royset, J., Bye, R., and Oppen, J. (2018). Preventing environmental disasters from grounding accidents: A case study of tugboat positioning along the norwegian coast. *Journal of the Operational Research Society*, 69(11):1773–1792.

Backe, S. and Haugland, D. (2017). Strategic optimization of offshore wind farm installation. *Lecture Notes in Computer Science*, 10572:285–299.

Bakker, H., Dunke, F., and Nickel, S. (2019). A structuring review on multi-stage optimization under uncertainty: Aligning concepts from theory and practice. *Omega*, 96(Article ID:102080).

© The Editor(s) (if applicable) and The Author(s), under exclusive license to Springer Fachmedien Wiesbaden GmbH, part of Springer Nature 2026

D. Rippel, *Online Decision Support for Offshore Wind Farm Installations*, https://doi.org/10.1007/978-3-658-49912-9

Baldea, M., Du, J., Park, J., and Harjunkoski, I. (2015). Integrated production scheduling and model predictive control of continuous processes. *AIChE Journal*, 61(12):4179–4190.

Barlow, E., Tezcaner Öztürk, D., Revie, M., Akartunali, K., Day, A., and Boulougouris, E. (2018). A mixed-method optimisation and simulation framework for supporting logistical decisions during offshore wind farm installations. *European Journal of Operational Research*, 264(3):894–906.

Bartolucci, L., Cordiner, S., Mulone, V., Santarelli, M., Lombardi, P., Arendarski, B., and Komarnicki, P. (2019). MPC-based electric energy storage sizing for a net zero energy factory.

Beinke, T. (2018). *Kooperative Errichtung der Offshore-Windenergie—Konzeptionelle Entwicklung einer lieferketten- und projektübergreifenden Logistikgestaltung*. PhD thesis, Universität Bremen.

Beinke, T., Ait Alla, A., and Freitag, M. (2017). Resource sharing in the logistics of the offshore wind farm installation process based on a simulation study. *International Journal of e-Navigation and Maritime Economy*, 7(6):42–54.

Beinke, T., Ait Alla, A., Oelker, S., and Freitag, M. (2020a). Demand for special vessels for the decommissioning of offshore wind turbines in the german north sea—a simulation study. In Chung, J. S., Akselsen, O. M., Jin, H., H., K., Lee, Y., and Matskevitch, D., editors, *Proceedings of the 30th International Ocean and Polar Engineering Conference (ISOPE)*, pages 480–484. ISOPE:Cupertino.

Beinke, T., Quandt, M., Ait Alla, A., and Freitag, M. (2020b). The impact of information sharing on installation processes of offshore wind farms—process modelling and simulation based analysis. *Int. J. of Shipping and Transport Logistics (IJSTL)*, 12(1):117–145.

Borthen, T., Loennechen, H., Fagerholt, K., Wang, X., and Vidal, T. (2019). Bi-objective offshore supply vessel planning with costs and persistence objectives. *Computers and Operations Research*, 111:285–296.

Breton, S. and Moe, G. (2009). Status, plans and technologies for offshore wind turbines in europe and north america. *Renewable Energy*, 34(3):646–654.

BVGassociates (2010). Guide to an offshore wind farm. https://bvgassociates.com/publications/, last accessed 08.04.2020.

BVGassociates (2019). Guide to an offshore wind farm. https://bvgassociates.com/publications/, last accessed 08.04.2020.

Cai, H., Li, P., Su, C., and Cao, J. (2018). Double-layered nonlinear model predictive control based on hammerstein–wiener model with disturbance rejection. *Measurement and Control*, 51(7–8):260–275.

Cataldo, A. and Scattolini, R. (2014). Modeling and model predictive control of a demanufacturing plant. In *IEEE Conference on Control Applications (CCA), Juan Les Antibes, France, October 8–10*, pages 1855–1860.

Chaudhrya, I. A. and Khanb, A. A. (2016). A research survey: Review of flexible job shop scheduling techniques. *International Transactions in Operational Research*, 23:551–591.

Cheng, M.-Y., Wu, Y.-F., Wu, Y.-W., and Ndure, S. (2019). Fuzzy bayesian schedule risk network for offshore wind turbine installation. *Ocean Engineering*, 188(Article ID: 106238):1–19.

Chu, Y. and You, F. (2014). Moving horizon approach of integrating scheduling and control for sequential batch processes. *AIChE Journal*, 60(5):1654–1671.

Daenzer, W. F. and Haberfellner, R. (2002). *Systems Engineering: Methodik und Praxis*. Industrielle Organisation: Zürich, 11th edition edition.

Dalgic, Y., Lazakis, I., and Turan, O. (2013). Vessel charter rate estimation for offshore wind O&M activity. In Soares, C. G., editor, *Developments in maritime transportation and exploitation of sea resources : Proceedings of IMAM 2013, 15th International Congress of the International Maritime Association of the Mediterranean (IMAM)*, pages 899–907. CRC Press: Boca Raton.

Dastyar, H., Rippel, D., and Freitag, M. (2020a). Optimization of supplier development under market dynamics. *Mathematical Problems in Engineering*, 2020(Article ID: 2912380):1–18.

Dastyar, H., Rippel, D., Pannek, J., Thoben, K.-D., and Freitag, M. (2020b). A numerical study on the effects of trust in supplier development. *Processes*, 8(3):1–22.

Demir, Y. and İşleyen, K. (2013). Evaluation of mathematical models for flexible job-shop scheduling problems. *Applied Mathematical Modelling*, 37:977–988.

Deutsche WindGuard GmbH (2019). Status des Offshore-Windenergieausbaus in Deutschland—Erstes Halbjahr 2019. https://www.wind-energie.de/fileadmin/redaktion/dokumente/pressemitteilungen/2019/Status_des_Offshore-Windenergieausbaus_Halbjahr_2019.pdf, last accessed 23.01.2020.

Deutscher Wetterdienst (2020). Verification of results of numerical weather prediction. https://www.dwd.de/EN/research/weatherforecasting/num_modelling/05_verification/verification_node.html, last accessed 12.08.2020.

Devoy McAuliffe, F., Lynch, K., Sperstad, I., Nonas, L., Halvorsen-Weare, E., Jones, D., Akbari, N., Wall, G., Irawan, C., Norstad, I., Stalhane, M., and Murphy, J. (2018). The leanwind suite of logistics optimisation and full lifecycle simulation models for offshore wind farms. In *Journal of Physics: Conference Series*, volume 1104, pages 1–12.

Dewan, A., Asgarpour, M., and Savenije, R. (2015). *Commercial Proof of Innovative Offshore Wind Installation Concepts Using Ecn Install Tool*. ECN: Petten.

Dias, L. S. and Ierapetritou, M. G. (2016). Integration of scheduling and control underuncertainties: Review and challenges. *Chemical Engineering Research and Design*, 116:98–113.

Dulcic, Z., Pavlic, D., and Silic, I. (2012). Evaluating the intended use of decision support system (dss) by applying technology acceptance model (tam) in business organizations in croatia. *Procedia—Social and Behavioral Sciences*, 58:1565–1575.

Durakovic, A. (2020). Offshore wind faces installation vessel shortage—Rystad Energy. Online Magazine: offshoreWIND.biz. https://www.offshorewind.biz/2020/11/26/offshore-wind-faces-installation-vessel-shortage-rystad-energy/, last access 12.01.2021.

EEG (2021). Gesetz für den Ausbau erneuerbarer Energien (Erneuerbare-Energien-Gesetz). https://www.gesetze-im-internet.de/eeg_2014/BJNR106610014.html, last access: 12.01.2021.

Endrerud, O.-E., Liyanage, J., and Keseric, N. (2014). Marine logistics decision support for operation and maintenance of offshore wind parks with a multi method simulation model. In Tolk, A., Diallo, S. Y., Ryzhov, I. O., Yilmaz, L., Buckley, S., and Miller, J. A., editors, *Proceedings of the Winter Simulation Conference 2014*, pages 1712–1722. IEEE: New York.

EUROGATE, C. (2019). Preis- und Konditionsverzeichnis, 01. November 2019 (gültig bis 31.12.2020). http://www.eurogate.corpex-kunden.de/de/Terminals/Wilhelmshaven, last accessed: 26.06.2020.

Fachagentur Windenergie an Land e. V. (2020). Überblick zu den Abstandsempfehlungen zur Ausweisung von Windenergiegebieten in den Bundesländern (Stand 02.2020). https://www.fachagentur-windenergie.de/fileadmin/files/PlanungGenehmigung/FA_Wind_Abstandsempfehlungen_Laender.pdf, last accessed 24.03.2020.

Fred. Olsen Windcarrier (2020). List of jack-up vessels. https://windcarrier.com/fleet, last accessed 26.06.2020.

Gallestey, E., Stothert, A., Antoine, M., and Morton, S. (2002). Model predictive control and the optimization of power plant load while considering lifetime consumption. *IEEE Transactions on Power Systems*, 17(1):186–191.

Gambino, G., Verrilli, F., Del Vecchio, C., and Glielmo, L. (2016). Optimal operations and load allocation of a power plant equipped with a cchp feeding power, steam and cold water to an industrial plant. In *Proceedings of the European Control Conference (ECC'16)*, pages 328–333. IEEE:New York.

Garey, M. R., Johnson, D. S., and Sethi, R. (1976). The complexity of flowshop and jobshop scheduling. *Mathematics of Operations Research*, 1(2):117–129.

Gruber, J. and Prodanovic, M. (2014). Two-stage optimization for building energy management. *Energy Procedia*, 62:346–354.

Grüne, L. and Pannek, J. (2017). *Nonlinear Model Predictive Control—Theory and Algorithms*. Springer International Publishing: Cham, 2nd edition edition.

Gupta, D. and Maravelias, C. (2017). A general state-space formulation for online scheduling. *Processes*, 5(4). ArticleID: 69.

Hazir, Ö. (2015). A review of analytical models, approaches and decision support tools in project monitoring and control. *International Journal of Project Management*, 33(4):808–815.

Herroelen, W. and Leus, R. (2007). Robust and reactive project scheduling: A review and classification of procedures. *International Journal of Production Research*, 42(8):1599–1620.

Hevner, A. R. (2007). A three cycle view of design science research. *Scandinavian Journal of Information Systems*, 19(2):1–6.

Hillier, F. and Lieberman, G. (2001). *Introduction to operations research*. McGraw-Hill: Boston, 7th edition edition.

Holjevac, N., Capuder, T., Kuzle, I., Zhang, N., and Kang, C. (2018). Modelling aspects of flexible multi-energy microgrids. In *Proceedings of the 20th Power Systems Computation Conference*, pages 1–7. IEEE: New York.

Irawan, C. A., Wall, G., and Jones, D. (2017). Bi-objective optimisation model for installation scheduling in offshore wind farms. *Computers & Operations Research*, 78:393–407.

Irawan, C. A., Wall, G., and Jones, D. (2019). An optimisation model for scheduling the decommissioning of an offshore wind farm. *OR Spectrum*, 41(3):513 548.

Ivanov, D., Dolgui, A., and Sokolov, B. (2012). Applicability of optimal control theory to adaptive supply chain planning and scheduling. *Annual Reviews in Control*, 36(1):73–84.

Ivanov, D., Sethi, S., Dolgui, A., and Sokolov, B. (2018). A survey on control theory applications to operational systems, supply chain management, and industry 4.0. *Annual Reviews in Control*, 46:134–147.

Ji, J., Khajepour, A., Melek, W. W., and Huang, Y. (2017). Path planning and tracking for vehicle collision avoidance based on model predictive control with multiconstraints. *IEEE Transactions on Vehicular Technology*, 66(2):952–964.

Jung, T., Jang, H., and Lee, J. (2015). Move blocking strategy applied to re-entrant manufacturing line scheduling. *International Journal of Control, Automation and Systems*, 13(2):410–418.

Kahani, N., Bagherzadeh, M., Cordy, J. R., Dingel, J., and Varró, D. (2019). Survey and classification of model transformation tools. *Software & Systems Modeling*, 18(4):2361–2397.

Kerkhove, L.-P. and Vanhoucke, M. (2017). Optimised scheduling for weather sensitive offshore construction projects. *Omega*, 66:58–78.

Knudsen, B., Whitson, C., and Foss, B. (2014). Shale-gas scheduling for natural-gas supply in electric power production. *Energy*, 78:165–182.

Krallmann, H., Bobrik, A., and Levina, O. (2013). *Systemanalyse im Unternehmen: Prozessorientierte Methoden der Wirtschaftsinformatik*. Oldenbourg Verlag: München, 6th edition edition.

Lange, K., Rinne, A., and Haasis, H.-D. (2012). Planning maritime logistics concepts for offshore wind farms: A newly developed decision support system. *Lecture Notes in Computer Science: Computational Logistics*, 7555:142–158.

Lefebvre, D. (2018). Near-optimal scheduling for petri net models with forbidden markings. *IEEE Transactions on Automatic Control*, 63(8):2550–2557.

Leontaris, G., Morales-Nápoles, O., Dewan, A., and Wolfert, A. (2019). Decision support for offshore asset construction using expert judgments for supply disruptions risk. *Automation in Construction*, 107(Article ID: 102903).

Leontaris, G., Morales-Nápoles, O., and Wolfert, A. (2017). Planning cable installation activities for offshore wind farms including risk of supply delays. In Walls, L., Revie, M., and Bedford, T., editors, *Risk, Reliability and Safety: Innovating Theory and Practice*, pages 660–666. Taylor & Francis: London.

Lin, X., Negenborn, R., and Lodewijks, G. (2018). Predictive quality-aware control for scheduling of potato starch production. *Computers and Electronics in Agriculture*, 150:266–278.

Lindholm, A., Giselsson, P., Quttineh, N.-H., Lidestam, H., Johnsson, C., and Forsman, K. (2013). Production scheduling in the process industry. In *22nd International Conference on Production Research, Iguassu Falls, Paraná, Brasil, July 28—August 01*.

McCarthy, A., Hancock, N., and Raine, S. (2014). Simulation of irrigation control strategies for cotton using model predictive control within the variwise simulation framework. *Computers and Electronics in Agriculture*, 101:135–147.

Meyer, M. (2014). Reeder von offshore-spezialtonnage müssen spekulativer vorgehen. *HANSA International Maritime Journal*, 151(10):54–55.

Mirakhorli, A. and Dong, B. (2016). Occupancy behavior based model predictive control for building indoor climate—a critical review. *Energy and Buildings*, 129:499–513.

Mouli, G., Kefayati, M., Baldick, R., and Bauer, P. (2019). Integrated pv charging of ev fleet based on energy prices, v2g, and offer of reserves. *IEEE Transactions on Smart Grid*, 10(2):1313–1325.

Mrugalska, B. and Stetter, R. (2019). Health-aware model-predictive control of a cooperative agv-based production system. *Sensors*, 19(Article ID: 532):1–27.

Muhabie, Y. T., Rigo, P., Cepeda, M., and D'Agosto, M. A. (2018). A discrete-event simulation approach to evaluate the effect of stochastic parameters on offshore wind farms assembly strategies. *Ocean Engineering*, 149:279–290.

Object Management Group (2020). MDA Overview. https://www.omg.org/mda/specs.htm, last access 18.05.2020.

Oelker, S., Ait Alla, A., Büsing, S., Lütjen, M., and Freitag, M. (2020). Simulative approach for the optimization of logistic processes in offshore ports. In Chung, J. S., Akselsen, O. M., Jin, H., H., K., Lee, Y., and Matskevitch, D., editors, *Proceedings of the 30th International Ocean and Polar Engineering Conference (ISOPE)*, pages 446–452. ISOPE:Cupertino.

Oelker, S., Ait Alla, A., Lütjen, M., Lewandowski, M., Freitag, M., and Thoben, K.-D. (2018). A simulation study of feeder-based installation concepts for offshore wind farms. In Chung, J. S., Hyun, B.-S., Matskevitch, D., and Wang, A. M., editors, *Proceedings of the Twenty-eighth (2018) International Ocean and Polar Engineering Conference (ISOPE)*, pages 578–583. ISOPE:Cupertino.

Ouelhadj, D. and Petrovic, S. (2009). A survey of dynamic scheduling in manufacturing systems. *Journal of Scheduling*, 12:417–431.

PBT Netherlands BV (2020). Oil price information. https://www.petrolbunkering.com/price-information/, last accessed 26.06.2020.

Peffers, K., Tuunanen, T., Rothenberger, M. A., and Chatterjee, S. (2007). A design science research methodology for information systems research. *Journal of Management Information Systems*, 24(3):45–77.

Peng, S., Rippel, D., Lütjen, M., Becker, M., and Szczerbicka, H. (2020a). Simulation-based scheduling for offshore wind farm installation using timed petri nets approach. In *Proceedings of the 2020 Summer Simulation Conference*, volume 52 of *Simulation Series*, pages 56–67. Society for Computer Simulation International: San Diego.

Peng, S., Szczerbicka, H., and Becker, M. (2020b). Modeling and simulation of offshore wind farm installation with multi-leveled cgspn. In Chung, J. S., Akselsen, O. M., Jin, H., H., K., Lee, Y., and Matskevitch, D., editors, *Proceedings of the 30th International Ocean and Polar Engineering Conference (ISOPE)*, pages 472–479. ISOPE:Cupertino.

Ponsignon, T. and Mönch, L. (2014). Simulation-based performance assessment of master planning approaches in semiconductor manufacturing. *Omega*, 46:21–35.

Power, D. J. and Sharda, R. (2007). Model-driven decision support systems: Concepts and research directions. *Decision Support Systems*, 43(3):1044–1061.

Quandt, M., Beinke, T., Ait Alla, A., and Freitag, M. (2017). Simulation based investigation of the impact of information sharing on the offshore wind farm installation process. *Journal of Renewable Energy*, 2017:11 pages.

Raković, S. V. and Levine, W. S., editors (2019). *Handbook of Model Predictive Control*. Springer International Publishing, Cham.

Rawlings, J. B., Angeli, D., and Bates, C. N. (2012). Fundamentals of economic model predictive control. In *Proceedings of the 51st IEEE Conference on Decision and Control (CDC)*, pages 3851–3861. IEEE: New York.

REN21 (2019). Renewables 2018 global status report. Technical report, REN21. ISBN 978-3-9818911-3-3, https://www.energia.org/renewables-2018-global-status-report-ren21, last accessed 07.04.2020.

REN21 (2020). Renewables 2019 global status report. Technical report, REN21. ISBN 978-3-9818911-7-1, https://wedocs.unep.org/handle/20.500.11822/28496, last accessed 07.04.2020.

Rippel, D., Jathe, N., Becker, M., Lütjen, M., Szczerbicka, H., and Freitag, M. (2019a). A Review on the Planning Problem for the Installation of Offshore Wind Farms. *IFAC-PapersOnLine*, 52(13):1337–1342.

Rippel, D., Jathe, N., Lütjen, M., and Freitag, M. (2019b). Evaluation of loading bay restrictions for the installation of offshore wind farms using a combination of mixed-integer linear programming and model predictive control. *Applied Sciences*, 9(23):ArticleID: 5030.

Rippel, D., Jathe, N., Lütjen, M., and Freitag, M. (2020a). Base-port capacity optimization for the installation of offshore wind farms. In *Proceedings of the 2020 Summer Simulation Conference (SummerSim'20)*, volume 52 of *Simulation Series*, pages 32–43. Society for Computer Simulation International: San Diego.

Rippel, D., Jathe, N., Lütjen, M., and Freitag, M. (2020b). A mixed-integer formulation to optimize the resupply of components for the installation of offshore wind farms. In Bae, K.-H., Feng, B., Kim, S., Lazarova-Molnar, S., Zheng, Z., Roeder, T., and Thiesing, R., editors, *Proceedings of the 2020 Winter Simulation Conference (WSC'20)*, pages 1420–1431. IEEE: New York.

Rippel, D., Jathe, N., Lütjen, M., Szczerbicka, H., and Freitag, M. (2019c). Integrated domain model for operative offshore installation planning. In Ringle, C. M., Kersten, W., and Jahn, C., editors, *Digital Transformation in Maritime and City Logistics*, Proceedings of the Hamburg International Conference of Logistics (HICL), pages 25–54. epubli GmbH: Berlin.

Rippel, D., Jathe, N., Lütjen, M., Szczerbicka, H., and Freitag, M. (2019d). Simulation and optimization of operations for offshore installations planning using a model predictive control scheme. In Mustafee, N., Bae, K.-H., Lazarova-Molnar, S., Rabe, M., Szabo, C., Haas, P., and Son, Y.-J., editors, *Proceedings of the 2019 Winter Simulation Conference (WSC'19)*, pages 1719–1730. IEEE: New York.

Rippel, D., Peng, S., Lütjen, M., Szczerbicka, H., and Freitag, M. (2020c). Model transformation framework for scheduling offshore logistics. In Jahn, C., Kersten, W., and Ringle, C. M., editors, *Data Science in Maritime and City Logistics*, volume 30 of *Proceedings of the Hamburg International Conference of Logistics (HICL)*, pages 521–552. epubli GmbH: Berlin.

Rossi, F., Casas-Orozco, D., Reklaitis, G., Manenti, F., and Buzzi-Ferraris, G. (2017). A computational framework for integrating campaign scheduling, dynamic optimization and optimal control in multi-unit batch processes. *Computers and Chemical Engineering*, 107:184–220.

Sahin, F., Narayanan, A., and Robinson, E. P. (2013). Rolling horizon planning in supply chains: Review, implications and directions for future research. *International Journal of Production Research*, 51(18):5413–5436.

Sarker, B. R. and Faiz, T. I. (2017). Minimizing transportation and installation costs for turbines in offshore wind farms. *Renewable Energy*, 101:667–679.

Scholz-Reiter, B., Karimi, H. R., Lütjen, M., Heger, J., and Schweizer, A. (2011). Towards a heuristic for scheduling offshore installation processes. In Maneesh, S., Rao, R., and Liyanage, J. P., editors, *Proceedings of the 24th International Congress on Condition Monitoringand and Diagnostics Engineering Management. Advances in Industrial Integrated Asset Management*, pages 999–1008. COMADEM International: Birmingham.

Scholz-Reiter, B., Lütjen, M., Heger, J., and Schweizer (Virnich), A. (2010). Planning and control of logistics for offshore wind farms. In Rogozea, L., editor, *12th International*

Conference on Mathematical and Compuational Methods in Science and Engineering (MACMESE'10), Faro, Portugal, November 03–05, pages 242–247.

Schweizer, A., Beinke, T., and Scholz-Reiter, B. (2011). Das logistiknetzwerk der offshore-windenergie. *Industrie Management*, 27(6):13–16.

Shehzad, M., Abdelghany, M., Liuzza, D., and Glielmo, L. (2019). Modeling of a hydrogen storage wind plant for model predictive control management strategies. In *Proceedings of the 18th European Control Conference (ECC)*, pages 1896–1901. IEEE: New York.

Shim, J., Warkentin, M., Courtney, J. F., Power, D. J., Sharda, R., and Carlsson, C. (2002). Past, present, and future of decision support technology. *Decision Support Systems*, 33(2):111–126.

Sokoler, L., DInesen, P., and Jorgensen, J. (2017). A hierarchical algorithm for integrated scheduling and control with applications to power systems. *IEEE Transactions on Control Systems Technology*, 25(2):590–599.

Sperstad, I., Stålhane, M., Dinwoodie, I., Endrerud, O.-E., Martin, R., and Warner, E. (2017). Testing the robustness of optimal access vessel fleet selection for operation and maintenance of offshore wind farms. *Ocean Engineering*, 145:334–343.

Sprodowski, T., Sagawa, J. K., and Pannek, J. (2018). Frequency based model predictive control of a manufacturing system. *IFAC-PapersOnLine*, 51(2):801–806.

Statistisches Bundesamt Deutschland (2020). Daten zur Energiepreisentwicklung - Lange Reihen bis Mai 2020. https://www.destatis.de/DE/Themen/Wirtschaft/Preise/Publikationen/Energiepreise/energiepreisentwicklung-pdf-5619001.html, last accessed 26.06.2020.

Stiftung OFFSHORE-WINDENERGIE (2020). Grafik Ausbaustand Offshore-Windenergie Deutschland 2020 (Stand 30.6.2020). https://www.offshore-stiftung.de/sites/offshorelink.de/files/mediaimages/SOW_Offshorewind%20U%CC%88bersicht_DE_100720.jpg, last accessed 25.01.2021.

Stuckenschmidt, H. (2011). *Ontologien—Konzepte, Technologien und Anwendungen*. Springer International Publishing: Cham.

Sun, X., Huang, D., and Wu, G. (2012). The current state of offshore wind energy technology development. *Energy*, 41:298–312.

Tezcaner Öztürk, D., Barlow, E., Akartunali, K., Revie, M., Day, A., and Boulougouris, E. (2017). Robust rolling horizon optimisation model for offshore wind farm installation logistics. Workingpaper, University of Strathclyde. online at https://pureportal.strath.ac.uk/en/publications/robust-rolling-horizon-optimisation-model-for-offshore-wind-farm-, last accessed 13.02.2021.

Tong, C., Palazoglu, A., and El-Farra, N. (2017). A decomposition scheme for integration of production scheduling and control: Demand response to varying electricity prices. *Industrial and Engineering Chemistry Research*, 56(31):8917–8926.

Ursavas, E. (2017). A benders decomposition approach for solving the offshore wind farm installation planning at the north sea. *European Journal of Operational Research*, 258:703–714.

Vargas-Villamil, F. and Rivera, D. (2000). Multilayer optimization and scheduling using model predictive control: Application to reentrant semiconductor manufacturing lines. *Computers and Chemical Engineering*, 24(8):2009–2021.

Vis, I. F. and Ursavas, E. (2016). Assessment approaches to logistics for offshore wind energy installation. *Sustainable Energy Technologies and Assessments*, 14:80–91.

Wang, X. and Kopfer, H. (2015). Rolling horizon planning for a dynamic collaborative routing problem with full-truckload pickup and delivery requests. *Flexible Services and Manufacturing Journal*, 27(4):509–533.

Wenzelburger, P. and Allgöwer, F. (2019). A novel optimal online scheduling scheme for flexible manufacturing systems. *IFAC-PapersOnLine*, 52(10):1–6.

Whiteman, A., Rueda, S., Akande, D., Elhassan, N., Escamilla, G., and Arkhipova, I. (2020). Renewable capacity statistics 2020. Technical report, IRENA - International Renewable Energy Agency. https://www.irena.org/publications/2020/Mar/Renewable-Capacity-Statistics-2020, last accessed 07.04.2020.

WindSeeG (2020). Gesetz zur Entwicklung und Förderung der Windenergie auf See (Windenergie-auf-See-Gesetz. https://www.gesetze-im-internet.de/windseeg/BJNR231000016.html, last access 12.01.2021.

Wiser, R. and Bolinger, M. (2018). 2018 wind technologies market report. Technical report, U.S. Department of Energy. https://www.energy.gov/eere/wind/downloads/2018-wind-technologies-market-report, last accessed 07.04.2020.

Wojakowski, P. and Warzolek, D. (2014). The classification of scheduling problems under production uncertainty. *Research in Logistics & Production*, 4(3):245–255.

Zhang, J., Ding, G., Zou, Y., Qin, S., and Fu, J. (2019a). Review of job shop scheduling research and its new perspectives under industry 4.0. *Journal of Intelligent Manufacturing*, 30:1809–1830.

Zhang, Q., Liu, P., and Pannek, J. (2019b). Combining mpc and integer operators for capacity adjustment in job-shop systems with rmts. *International Journal of Production Research*, 57(8):2498–2513.

Zheng, H.-Y., Zeng, J., Zou, T., Hu, J.-T., and Yu, H.-B. (2017). A multi-variable real-time online decision support system with multi-priority. *Control Theory and Applications*, 34(12):1605–1613.

Ziarnetzky, T., Mönch, L., and Uzsoy, R. (2018). Rolling horizon, multi-product production planning with chance constraints and forecast evolution for wafer fabs. *International Journal of Production Research*, 56(18):6112–6134.

MIX
Papier aus verantwortungsvollen Quellen
Paper from responsible sources
FSC® C105338

If you have any concerns about our products,
you can contact us on
ProductSafety@springernature.com

In case Publisher is established outside the EU,
the EU authorized representative is:
Springer Nature Customer Service Center GmbH
Europaplatz 3, 69115 Heidelberg, Germany

Printed by Libri Plureos GmbH
in Hamburg, Germany